THE METAPHYSICS WITHIN PHYSICS

What fundamental account of the world is implicit in physical theory? Physics straightforwardly postulates quarks and electrons, but what of the more intangible elements, such as laws of nature, universals, causation and the direction of time? Do they have a place in the physical structure of the world?

Tim Maudlin argues that the ontology derived from physics takes a form quite different from those most commonly defended by philosophers. Physics postulates irreducible fundamental laws, eschews universals, does not require a fundamental notion of causation, and makes room for the passage of time. In a series of linked essays *The Metaphysics Within Physics* outlines an approach to metaphysics opposed to the Humean reductionism that motivates much analytical metaphysics.

Tim Maudlin is Professor of Philosophy at Rutgers, The State University of New Jersey.

The Metaphysics Within Physics

TIM MAUDLIN

OXFORD
UNIVERSITY PRESS

Great Clarendon Street, Oxford OX2 6DP

Oxford University Press is a department of the University of Oxford.
It furthers the University's objective of excellence in research, scholarship,
and education by publishing worldwide in

Oxford New York

Auckland Cape Town Dar es Salaam Hong Kong Karachi
Kuala Lumpur Madrid Melbourne Mexico City Nairobi
New Delhi Shanghai Taipei Toronto

With offices in

Argentina Austria Brazil Chile Czech Republic France Greece
Guatemala Hungary Italy Japan Poland Portugal Singapore
South Korea Switzerland Thailand Turkey Ukraine Vietnam

Oxford is a registered trade mark of Oxford University Press
in the UK and in certain other countries

Published in the United States
by Oxford University Press Inc., New York

© Tim Maudlin 2007

The moral rights of the author have been asserted
Database right Oxford University Press (maker)

First published 2007
First published in paperback 2010

All rights reserved. No part of this publication may be reproduced,
stored in a retrieval system, or transmitted, in any form or by any means,
without the prior permission in writing of Oxford University Press,
or as expressly permitted by law, or under terms agreed with the appropriate
reprographics rights organization. Enquiries concerning reproduction
outside the scope of the above should be sent to the Rights Department,
Oxford University Press, at the address above

You must not circulate this book in any other binding or cover
and you must impose the same condition on any acquirer

British Library Cataloguing in Publication Data
Data available

Library of Congress Cataloging in Publication Data
Data available

Typeset by Laserwords Private Limited, Chennai, India
Printed in the UK
by
MPG Books Group

ISBN 978–0–19–921821–9 (Hbk.); 978–0–19–957537–4 (Pbk.)

1 3 5 7 9 10 8 6 4 2

For Clio and Maxwell

Acknowledgements

Reflecting on the chances and vicissitudes of life, I am constantly struck by how undeservedly lucky I have been. Without that luck, in its many forms, this volume would not exist.

I have been immeasurably fortunate to have sharp, stimulating, patient, encouraging colleagues and friends. The positions defended in these papers have arisen out of lively interchange with many people, but foremost among them, over many years, have been David Albert, Frank Arntzenius, Shelly Goldstein, Ned Hall, and Barry Loewer. Some have sometimes been opponents, some sometimes allies, and all (at various times) simply puzzled, and I have learned immensely from all of them. If I hadn't found myself in their company I would not be who I am, either as a philosopher or as a human being.

As they all know, and I discover anew every day, I have been unspeakably lucky in my family. My wife Vishnya has lived with these ideas, and played midwife to them, over the last decade and a half. They only exist because we have been able to nourish them, and criticize them, and correct them, together.

Through the same time, we had the delight and joy to see our children, Clio and Maxwell, grow and thrive. It is a standard trope to ask one's children's understanding for the time that the composition of a book has taken from them, but I have no such apologies to make. These papers were mostly written late at night, after they were in bed, because it was more important, and more inspiring, and more fun, to spend the days with them. Perhaps one day they will become curious about what their father was up to in the wee hours, and will turn to this volume to find out. However that may be, I dedicate it to them with my gratitude and love.

T.M.

Contents

Introduction	1
1. A Modest Proposal Concerning Laws, Counterfactuals, and Explanations	5
2. Why Be Humean?	50
3. Suggestions from Physics for Deep Metaphysics	78
4. On the Passing of Time	104
5. Causation, Counterfactuals, and the Third Factor	143
6. The Whole Ball of Wax	170
Epilogue: A Remark on the Method of Metaphysics	184
References	192
Index	195

Introduction

The essays that compose this book were written over a span of more than a decade, and were not originally conceived of as part of a larger philosophical project. But it transpires that they collaborate: all have been fashioned from the same clay and molded by the same concerns. The basic idea is simple: metaphysics, insofar as it is concerned with the natural world, can do no better than to reflect on physics. Physical theories provide us with the best handle we have on what there is, and the philosopher's proper task is the interpretation and elucidation of those theories. In particular, when choosing the fundamental posits of one's ontology, one must look to scientific practice rather than to philosophical prejudice.

From this point of view, a distressing amount of philosophical energy appears to be invested in questionable projects. For example, it has been a long-standing philosophical problem to provide an 'analysis' or an 'account' or a 'reduction' of laws of nature in terms of something else, such as relations between universals or patterns of local quantities. But nothing in scientific practice suggests that there should be such an analysis (unlike, say, genes which are explicable in terms of underlying physio-chemical structure). The first essay argues simply this: laws of nature stand in no need of 'philosophical analysis'; they ought to be posited as ontological bedrock.

The most frequent objection to this 'primitivist' view of laws is that it stands in opposition to an influential metaphysical picture that David Lewis advocates and elaborates in many of his works. The metaphysical picture goes by the name of Humean Supervenience:

Humean supervenience is named in honor of the greater [*sic*] denier of necessary connections. It is the doctrine that all there is to the world is a vast mosaic of local matters of fact, just one little thing and then another. (But it is no part of the thesis that these local matters of fact are mental.) We have geometry: a system of external relations of spatio-temporal distance between points. Maybe points of spacetime itself, maybe point-sized bits of matter or aether fields, maybe both. And at those points we have local qualities: perfectly natural intrinsic properties which need nothing bigger than a point at which to be instantiated. For short: we have an arrangement of qualities. And that is all. All else supervenes on that. (Lewis 1986a, p. x)

Accepting Humean Supervenience severely constrains one's ontological resources and correspondingly poses a daunting set of metaphysical challenges. Given only a patterned set of local qualities arrayed through space-time, one must derive laws, causes, truth conditions for counterfactuals, a direction of time, dispositions, objective chances, and so on. Lewis set his hand to these projects, and many more have followed. There is work enough here to sustain a large cadre of philosophers for many generations.

The Humean project is very seductive: one is given a delimited set of resources and set the task of expressing truth conditions for some class of propositions in those terms. To win the game is to get the truth conditions to come out in a way that is, largely, intuitively correct. Proposed solutions can be counter-exampled, counter-examples can be reinterpreted, intuitions can be bartered off against each other. If a proposed analysis fails, there is always the hope that one more widget, one extra subordinate clause, can set things right again. No end of cleverness can be deployed both on offense and defense.

For all that, I think that the Humean project, as Lewis conceived it, is unjustifiable. Why think that all there is to the world is 'a vast mosaic of local matters of fact'? Why accept these strong constraints on one's ontology in the first place? I take up these questions in the second Chapter: 'Why Be Humean?'

And it is not just that the Humean picture is too impoverished, that it postulates less than there is. The metaphysical atoms it utilizes are instances of repeatable (or qualitatively identical) local properties. These are also elements of the basic ontology of non-Humeans such as David Armstrong. But modern gauge theories paint a different picture: they provide an account of the physical nature of the world that does not employ such properties. This novel approach to the problem of universals is the subject of Chapter 3, 'Suggestions from Physics for Deep Metaphysics'.

As the project of reducing natural laws provides employment for metaphysicians, so the game of analyzing the nature of the 'arrow of time' occupies philosophers of science. The problematics are very similar: on the one hand, laws are claimed to be nothing but patterns in the physical state of the world; on the other, the direction of time is supposed to be nothing but a matter of how physical contents are disposed across the space-time manifold.

And corresponding to primitivism about laws is a primitivist approach to the arrow of time: it is a fundamental, irreducible fact that time is directed. Chapter 4, 'On the Passing of Time', defends this view.

So among the first four essays we have the following conclusions: the Humean project is unjustified, in that both laws of nature and the direction of time require no analysis, and is misconceived, in that the atoms it employs do not correspond to present physical ontology. But I do not wish to become a primitivist about everything. In particular, physical theory does not employ a notion of causation at a fundamental level, so causal locutions are proper candidates for reduction. Some efforts in this direction, consonant with my preferred ontology, are made in Chapter 5, 'Causation, Counterfactuals, and the Third Factor'.

Scattered among these five papers, then, lies the outlines of an ontology based in physics. Only once the papers were written, however, did their joint import become clear to me. In particular, taking both the laws of nature (as laws of temporal evolution) and the direction of time as primitive allows one to produce a sort of causal explanation of the fundamental Humean entity: the Humean Mosaic, or the total physical state of the universe. These disparate threads are pulled together in Chapter 6, 'The Whole Ball of Wax'. A final brief reflection on method in metaphysics can be found in the Epilogue.

If there is one major topic related to these papers that deserves more extensive treatment, it is Ockham's Razor. The Razor, like Humean Supervenience, generates much employment for philosophers: the more parsimonious the ontology, it is said, the better. Why believe in irreducible laws of nature if some passable replacement can be found simply in the patterns in the mosaic? Why believe in an intrinsic direction to time if the gradient of entropy always points the same way? *Entia non sunt multiplicanda praeter necessitatem*, and the availability of a reduction obviates any necessity. Surely we should be seeking the slenderest basis on which to erect our ontology.

But it is not clear that the Razor can withstand much critical scrutiny. If by *necessitas* one means *logical* necessity, then the Razor will land us in solipsism. But if one means something milder—entities ought not to be multiplied *without good reason*—then the principle becomes a harmless bromide: nor should one's ontology be *reduced* without good reason. The Razor cannot be derived from a simple logical observation—that the subjective probability one assigns to the existence of one set of items must always decrease when one enlarges the set—since the Razor recommends *positive disbelief* in the additional ontology. Such disbelief engenders errors if the controversial items exist: if the universe has been profligate, then the Razor will lead us astray.

Why, then, has the Razor been so widely accepted? No doubt, in many cases it yields the correct conclusion: explanations that require elaborate conspiracies and coincidences are less often true than simpler alternatives. But this result can frequently be derived straight from confirmation theory: simpler theories are commonly *better confirmed* by the data than competitors with equants and epicycles.[1] Yet this does not mean that the theory with the smaller ontology is *always* better confirmed. And questions about how one confirms—or disconfirms—claims about the ontological status of natural laws or the direction of time are bound to be extremely contentious.

So rather than a general theory of justification for ontological commitment, I have produced only some case studies. These studies suggest that the Razor, and the accompanying mania for ontological reduction, is overrated. The concepts of the laws of nature and of the passage of time play central roles in our picture of the world, and the arguments that these can, or need to be, reduced to something else strike me as flimsy. If the ontology that arises most naturally from reflection on physics is too rich for Ockham or Hume or Lewis, then so much the worse for them. Let others subsist on the thin gruel of minimalist metaphysics: I'll take my ontology *mit Schlag*.

A final note on the structure of these papers. Each was written to stand alone, rather than as part of a larger work. So each can be read independently of the others, but there is a corresponding need for some repetition and redundancy among them. On the theory that it is easier to skip what is familiar than to retrieve what is not, I have left them as they are. To any hardy soul who soldiers through them all: my thanks and my apologies.

---

[1] Clark Glymour's bootstrapping confirmation theory, for example, has as a consequence that certain sorts of 'deoccamized' theories will be less well confirmed than the theories from which they are generated (Glymour 1980, pp. 143ff.).

1

A Modest Proposal Concerning Laws, Counterfactuals, and Explanations

1. METHODOLOGICAL PROLEGOMENON

Philosophical analyses may be pursued via a myriad of methods in service of as great a multitude of goals. Frequently the data upon which an analysis rests, and from which it receives its original inspiration, recount systematic connections between diverse realms of discourse or diverse sets of facts, events, actions, or objects. The aim of the project is elucidating the underlying logical, conceptual, or ontological structure that accounts for these connections. As an obvious example, John's beliefs about what Sarah knows covary systematically with his beliefs about what Sarah believes, about what Sarah has good evidence for, and about what is actually the case. We may explain these covariations by postulating that John at least tacitly adheres to the theory that knowledge is some species of justified true belief.

The results of such a preliminary investigation of correlations among beliefs may be put to various uses. If we choose to endorse John's theory we will simply assert that what Sarah believes, what Sarah has good evidence for, and what is true determine what she knows. We may endorse John's theory, as revealed by his inferences, but criticize his particular judgements. For example, John's inferences may imply that he takes knowledge to require infallible evidence and so, by his own lights, he should not ascribe knowledge to Sarah since her evidence is not conclusive. Or we may instead endorse John's judgements and recommend that he amend his inferences accordingly. And, of course, the inferences, particular judgements, and intuitions at issue may be our own.

This essay was written in 1989, but being too long for a journal and too short for a book only circulated informally. There are evident similarities to John Carroll's approach in his *Laws of Nature* (1994), and we have both been identified as primitivists about laws. I have not attempted a direct comparison between our views as it would not fit into the structure of the paper as originally conceived.

Considerable light may be provided simply by displaying such connections among different realms of discourse. But once embarked upon the voyage, the siren song of reduction is difficult to resist. If the connections can be codified as explicit definitions we can cast out some of the primitive notions with which we are laden, reducing our ontology or ideology. Such reductions may be sought in two quite distinct ways. On the one hand, one may launch the enterprise with some preferred set of concepts, properties, or objects whose philosophical credentials are supposed to be already established. Reduction to this set then validates some otherwise suspect entities. On the other hand, the direction and form of the reduction may await the result of the analysis. Once the inferences that connect various domains have been mapped, one set of items may be found to provide resources to account for some other set. Examples of attempts at the first sort of reduction abound, especially among the logical empiricists. Hans Reichenbach's insistence that purported facts about the geometry of space be parsed into claims about observable coincidences of physical objects plus analytical co-ordinative definitions may serve as a clinical example: the epistemological status of the observable phenomena and of the definitions confer legitimacy upon any further concepts constructed from them (Reichenbach 1958, pp. 10–24). The second approach may be illustrated by David Lewis's realism about possible worlds (Lewis 1986b). Possible worlds hardly recommend themselves on either epistemological or ontological grounds as desirable primitives. Lewis argues, rather, that constructions founded on the plurality of worlds explicate, regiment, and provide semantic grounding for claims about possibility, counterfactuals, propositional content, etc. If we want all these claims to have truth values, Lewis argues, we had best abandon our prejudices and admit possible worlds as primitives in our ontology.

Having sketched a rough taxonomy of philosophical projects, the present endeavor can now be situated. As befits a modest proposal, the direct aims of this enquiry are slight. The primary goal is an outline of the systematic connections between beliefs about laws of nature and a small assortment of other beliefs. The examination will be carried out predominantly in what may be called the conceptual mode, focusing on inferences so as to sidestep the deep problem of ontological commitment. We may liken this to examining John's tacit theory of knowledge without affirming whether or not anyone has any knowledge, whether there is any such thing. As an example of the difference between the conceptual and ontological levels, consider a connection between laws of nature and counterfactuals that has been widely noted: laws, it is said, 'support' counterfactual claims while accidental regularities do not. Such 'support' can be interpreted in two ways. At the conceptual level, it means that if one assents to the proposition that 'all Fs are Gs' is a law, then one will generally also accept that had *s* been an F it would also have been a G. This is a datum about belief formation. On the ontological level, such 'support' would rather represent a relation among objective facts: the law, as a constituent of nature itself, provides grounds for the truth of the counterfactual. One can accept the datum about belief formation but deny any ontological implications by rejecting counterfactuals as candidates for truth. So the result of the investigation may be construed as a conditional: If one wants to assign truth values to counterfactuals then one must also accept laws among the furniture of the world. If one assigns all of the discourse about laws and counterfactuals to the limbo of the non-fact-stating, still the patterns that govern people's willingness to mouth these sentences must be explained.

This enquiry shall not be of the prejudgemental sort. No presuppositions shall be made about the preferability of one sort of discourse to another. Nor should we assume that any reduction must eventuate. If one set of concepts emerges as logically primitive it is because the connections among notions are strong and asymmetrical, allowing some to be generated from others but not vice versa.

Let us begin by setting out the domains of discourse that will be our focus and by sketching their connections to assertions about laws of nature. These connections suffice to upset the most influential philosophical accounts of laws of nature.

2. LAWS, POSSIBILITIES, COUNTERFACTUALS, AND EXPLANATIONS

Beliefs about laws of nature undoubtedly influence and are influenced by any number of other sorts of beliefs. Of these, three classes are of particular interest: beliefs about possibilities, about counterfactuals, and about explanations. Some few examples may illustrate each of these.

A physicist who accepts Einstein's General Theory of Relativity will also believe that it is physically possible for a universe to be closed (to collapse in a Big Crunch) and possible for a universe to be open (continue expanding forever). This is especially evident since we don't yet know whether our own universe is open or closed, so empirical data are still needed to determine which possibility obtains. But even if the issue were settled, the laws of gravitation, as we understand them, admit both possibilities. Anyone who accepts Einstein's laws of gravitation, or Newton's laws of motion and gravitation, must admit the physical possibility of a solar system with six planets even if no such system actually exists. If one believes that the laws of nature governing some sort of event, say a coin flip, are irreducibly probabilistic (and that the outcomes of flips are independent of one another) then one must admit it to be physically possible for any sequence of heads and tails to result from a series of flips.

I take these sorts of inference to be manifest in the actual practice of science and to be intuitively compelling. Any account of the nature of physical laws should account for them.

One connection between laws and counterfactuals has already been noted. If one accepts the conductivity of copper as a law, or as a consequence of laws, then one will also accept, in a wide variety of circumstances, that had a particular piece of copper been subjected to a voltage differential, it would have conducted electricity. Such inferences are notoriously fragile, and in many circumstances counterfactuals seem to have no determinate truth value even though the relevant laws of nature are not contested. This stands in need of explanation. But any acceptable account of laws and of counterfactuals must illuminate the relation of support between the former and the latter where it exists.

Finally, a more amorphous connection is generally acknowledged to hold between laws and explanations. The covering law model, for all its deficiencies, testifies to the depth of this relationship. Coming to see particular events or phenomena as manifestations of laws of nature can provide an understanding of them that does not follow from recognizing them as instances of accidental generalizations. A full elucidation of this fact would require a complete theory of explanation, a task far beyond our scope. But the connection does provide one touchstone for accounts of laws. A law ought to be capable of playing some role in explaining the phenomena that are governed by or are manifestations of it. And a physical event or state or entity which is already explained in all its details by some set of physical laws cannot provide good grounds for appending to these laws new ones. 'We are to admit no more causes of natural things than such as are both true and sufficient to explain their appearances' is Newton's first Rule of Reasoning in Philosophy (Newton 1966, p. 398). Any account which disrespects accepted links between laws and explanations thereby loses some of its plausibility.

By way of illustration of these connections, consider a case brought by Bas van Fraassen against the sophisticated regularity account of laws. It is a case to be more fully discussed presently.

> To say that we have the concept of a law of nature must imply at least that we can mobilize intuitions to decide on proffered individual examples. Let us then consider a possible world in which all the best true theories, written in an appropriate sort of language, include the statement that all and only spheres are gold. To be concrete, let it be a world whose regularities are correctly described by Newton's mechanics plus law of gravitation, in which there are golden spheres moving in stable orbits about one another, and much smaller iron cubes lying on their surface, and nothing else. If I am now asked whether in that world, all golden objects are spherical because they must be spherical, I answer *No*. First of all it seems to me that there could have been little gold cubes among the iron ones, and secondly, that several of the golden spheres could (given slightly different initial conditions) have collided with each other and thus altered each other's shapes. (1989, pp. 46–7)

The intuitive bite of van Fraassen's example derives from the sorts of connections remarked above. The Newtonian laws of gravitation and motion (plus whatever laws are needed for there to be gold and iron) seem clearly to admit of the possibility of a world such as van Fraassen describes. The shapes and dispositions of the contents of such a world would be set as initial conditions; the stability of the orbits, persistence of the objects, and lack of collisions would then follow from the laws. This scenario might not actually be a physical possibility given that Newton's laws do not obtain in our world. But if we accept that such laws might obtain then the possibility of the laws brings in train the possibility of a concrete situation such as described. The connection between laws and possibilities is manifest.

The connection between laws and explanations also plays a part. Why should we accept that it is *not* a law in this world that something is a sphere just in case it is gold? It can't be, as van Fraassen insinuates, because we already know that different initial conditions would yield non-spherical gold or non-golden spheres. This is a *petitio principii*; if we accept that it is a law, we will not admit the initial conditions as possible. Rather it is because if we assume that Newton's laws are the only laws operating, the sphericality of the gold can be accounted for by initial conditions. Together the initial conditions and Newton's laws entail all of the particular facts about this world. So no new laws need be invoked. This does not prove that there *could not* be a further law about spheres and gold, only that our intuitions accept that there *need not* be. And if there need not be, then a regularity appearing in all the best theories need not be a law.

Van Fraassen's example and variations on it demonstrate that this sophisticated regularity account, associated with Mill, Ramsey, and Lewis, cannot capture the connections between realms of discourse noted above. An acceptable theory of laws must. In the section that follows, I shall argue that none of the main philosophical accounts of laws of nature meets this challenge. Nor can van Fraassen's view, which eschews laws altogether, make sense of actual scientific practice. Our examination will take actual scientific practice as a starting point and return to it in the end.

3. THE LOGICAL FORM OF LAWS OF NATURE

Most philosophical analyses of laws of nature proceed from the logical empiricist formulation of laws of nature, at least as an initial approximation. In that tradition, the logical skeleton of a law is $(x)(Fx \supset Gx)$. A further vague requirement is added that the predicates 'F' and 'G' must be purely qualitative, i.e. contain no ineliminable reference to named individuals. The addendum is added to save the account from total vacuity, for even 'John went to the store' can be pressed into the logical Procrustean bed if the language contains an individual term denoting John and contains identity. But the addendum fails entirely since merely accidental universal concurrences of qualitative properties are possible and since, given a natural stock of spatial relations, each individual can be uniquely denoted by a purely qualitative predicate if the world contains any spatial asymmetry. Despite these drawbacks the logical skeleton serves as a starting point or inspiration for more sophisticated views.

If constraints on the predicates are not sufficient to pick out laws of nature, resort must be made to some more controversial means. One possibility is appeal to modality by the addition of a modal operator: $\Box(x)(Fx \supset Gx)$. The regularity that arises from the operation of laws of nature is neither logical nor metaphysical necessity, so the box must be interpreted as nomic necessity. But appending a box to a sentence and calling it nomic necessity is only marginally superior to appending 'It is a law that ...', and can hardly be considered an informative analysis.

Perhaps the difficulty lies instead in the horseshoe. Material implication is notoriously weak, so some more metaphysically substantive connective may be of use. It is not just that $(x)(Fx \supset Gx)$ is necessary, it is that being an F necessitates being a G. According to the view developed by David Armstrong,[1] a law holds if being an F necessitates being a G in virtue of a relation between

---

[1] Similar views have been developed by Micheal Tooley (1977) and Fred Dretske (1977). I am using Armstrong as an exemplar.

the universals F-ness and G-ness. In Armstrong's formulation the law is symbolized as N(F,G)(a's being F, a's being G), from which the offending horseshoe has been eliminated (Armstrong 1983, p. 90). This elimination, though, immediately poses a new problem, for $(x)(Fx \supset Gx)$ is supposed to follow from the law. In the view of some, e.g. van Fraassen, the gap between a fact about relations of universals and a fact about relations among their instances cannot be closed (van Fraassen 1989, pp. 97 ff). Further, Armstrong's approach may be considered *ignotum per ignotus*: do we really have a firmer grasp of what a necessitation relation between universals is than we do of what a law is? This is especially troubling since the species of necessitation at issue must again be denominated *nomic* necessitation.

Armstrong's account, although eschewing $(x)(Fx \supset Gx)$ as any sort of law, is still naturally regarded as influenced by that skeleton. The formula directs our attention to a relation between the predicates 'F' and 'G', and the universality of laws, enshrined in the quantifier, implies the universality of the relation. It is not a far step to suppose that this universal relation among instances derives from a relation among the universals denoted by the predicates. Universality was the only feature of laws that the positivists could get a clean syntactic grip on, and it continues to influence even those who firmly reject the positivist view. Indeed, Armstrong is able to admit the possibility of purely local laws only by the dubious admission of universals that are not purely qualitative (Armstrong 1983, p. 100).

Given that the preliminary supposition about the logical form of laws can so strongly influence subsequent avenues of research, we ought to pause to ask whether $(x)(Fx \supset Gx)$ is really a useful place to begin. Does it contain features that laws lack? Does it ignore structure commonly found in laws? An appropriate place to begin is with some real scientific theories rather than with cooked examples of the 'All ravens are black' variety. Let us look at some laws without formal prejudices and see what we find.

The fundamental law of Newtonian mechanics, the mathematical consequence of Newton's first two laws, is $\mathbf{F} = m\mathbf{a}$ or $\mathbf{F} = m\ d^2\mathbf{x}/dt^2$ or, most precisely, $\mathbf{F} = m\ d(m\mathbf{v})/dt$. The fundamental law of non-relativistic quantum mechanics, Schrödinger's equation, is $i\hbar\ \partial/\partial t\ |\Psi> =\ H\ |\Psi>$. No doubt these can be tortured into a form similar to $(x)(Fx \supset Gx)$, but it is hard to see what the purpose of the exercise would be. What is most obvious about these laws is that they describe how the physical state of a system or particle evolves through time. The laws are generally presumed to be universally valid. But this is not a feature directly represented in the formulae, nor does it appear to be essential to their status as laws. It is not contradictory to assert, or does not at first glance seem to be, that the evolution of physical states of particles is governed by Newton's laws *around here*, or that it has been *for the last 10 billion years* (but not before that). John Wheeler has proposed that after the Big Crunch the universe will be 'reprocessed' probabilistically, with the constants of nature and of motion, the number of particles and mass of the universe being changed (Misner, Thorne, and Wheeler 1973, p. 1214). It is a small step to suggest that the very laws themselves may change. Nor does it sound absurd to suggest that the laws elsewhere in the universe may differ from those here, especially if the various regions are prevented from interacting by domain walls of some sort. There might be some meta-law governing these different laws, or there might not be. But the supposition that Schrödinger's equation describes the evolution of physical quantities only in this 'bounce' of the universe, between our Big Bang and Big Crunch, doesn't seem incompatible with describing it still as a law, albeit a parochial one. At least, such a discovery would not appreciably alter our assessment of possibilities, counterfactuals, and explanations in most contexts. Astrophysics, biology, chemistry, physiology, and everyday language would be unchanged by the discovery. Parochial laws are still laws.

The laws cited above, then, tell us how, at least for some period and in some region, physical states evolve through time. Standing alone they are incomplete, for we need principles for determining the forces in Newtonian mechanics and the Hamiltonian operator H in quantum theory. Newton's third law and law of gravitation supply part of this demand. But the principle of temporal change is the motor of the enterprise. Supplying a force function for electrical interactions, frictional forces, etc. yields instances of Newtonian mechanics. One can change the form of a force function but stay within the Newtonian regime. Changing the law of temporal evolution, though, constitutes a rejection of Newtonian mechanics. Similarly, Schrödinger's equation, without any further specification of the Hamiltonian operator, is considered a fundamental principle of quantum mechanics. The specification of the Hamiltonian is a task that changes from one physical context to another.

Let us call a proposed basic law that describes the evolution of physical states through time a Fundamental Law of Temporal Evolution (FLOTE). Other sciences and folk wisdom recognize generalizations about temporal development that may be regarded as more or less lawlike, such as the Hardy–Weinberg law of population genetics or the observation that certain plants grow towards sources of light. Let us denominate these simply Laws of Temporal Evolution (LOTEs), being lawlike insofar as they are accepted as supporting counterfactuals and as supplying explanations. LOTEs are happily acknowledged to admit of exceptions (e.g. if a nuclear explosion occurs nearby, the plant won't grow towards light sources). But they still are accepted as describing how things would go, at least approximately, under normal conditions. LOTEs are generally thought to be ontologically parasitic on FLOTEs: in our world the laws of population genetics describe temporal changes in the gene pool in part because the laws of physics allow the physical realizations of the genes to have the properties—such as persistence through time, ability to recombine, etc.—which population genetics postulates of them. There is no such inverse dependence of FLOTEs on LOTEs.

Beside FLOTEs there are the adjunct principles that are needed to fill out the FLOTEs in particular contexts, principles about the magnitudes of forces and the form of the Hamiltonian, or about the sorts of physical states that are allowable. Some of these, such as Newton's law of gravitation, are laws of coexistence; others, such as the superselection rules of quantum mechanics, are constraints that are not naturally construed as either laws of succession or of coexistence. Some so-called laws of coexistence, such as the ideal gas law $PV = nRT$, are better construed as consequences of laws of temporal evolution. $PV = nRT$ is true only of equilibrium states: if we rapidly increase the volume of container, the pressure of the gas inside ceases to be well defined until after the relaxation time characteristic of the system has elapsed. The gas evolves into a state satisfying the law, and remains in such a state only so long as it is in equilibrium. Sorting out the status of these various adjunct principles and consequences of law is a task requiring nice discriminations that is foreign to our present purpose.

What is clear is that scientific and commonsense explanations demand the postulation of (F)LOTEs and their adjunct principles. It is only barely possible to conceive of a world that displays no diachronic regularity at all, in which earlier states are not even probabilistically associated with later ones. No object could persist, so the world would consist in point events forming no pattern through time. In a Newtonian setting, there still might be laws of coexistence among these events; in a relativistic regime, where there is no preferred simultaneity relation, total chaos would reign. And this is on the generous assumption that the very notion of a spatio-temporal structure could be defined absent any laws of temporal evolution.

FLOTEs can be deterministic or irreducibly stochastic. If the former, then the specification of appropriate boundary conditions suffices to fix the physical state in an entire region; if the latter, boundary conditions together with the law only determine a collection of possible physical states with associated probabilities.

To sum up this section, a look at laws used in science reveals a basic sort of law, the Law of Temporal Evolution, that specifies how specified states of a system will or can evolve into other states. Other laws are adjuncts to these, having content only in association with a FLOTE or with a LOTE designed for a more abstract level of description of the physical state. Such LOTEs apply to our world in virtue of the operation of FLOTEs. These laws may or may not be universal; in principle they might govern only a limited region. In the remainder of this paper we will consider how the idea of a FLOTE and its adjunct principles can illuminate the connections among laws, possibilities, counterfactuals, and explanations.

4. THE MODEST PROPOSAL

We have provided a rough characterization of laws of temporal evolution and adjunct principles. Taken together they provide descriptions of the state of a system and a rule, deterministic or probabilistic, of how that state evolves through time. So far no sort of philosophical analysis of these laws, or of lawhood, has been advanced. The temporal career of the world displays all sorts of regularities, described by more or less complex mathematical functions, including the accidental regularities that brought the logical empiricist account to grief. To make things worse, I have defended the claim that the regularities due to law need not persist through all time or obtain in all places since the laws may not. So what makes some such regularities into laws of nature, or into the consequences of law?

This question can take two forms. On the ontological side we may seek some further fact or structure that confers lawhood. For Armstrong, to give an example, the relevant further structure is a relation between universals. On the epistemological side, we may ask how we know which observed regularities are consequences of laws as opposed to being accidental. Armstrong must

A Modest Proposal

adopt at least some degree of skepticism here since no *observable* relations among instances of universals guarantee that there exists the relation of necessitation among the universals themselves. Even an ideal observer who sees everything that can be seen in the whole history of the universe cannot be entirely confident to have gotten the laws right.

On Lewis's sophisticated regularity view[2] the two questions have a single solution. What makes a regularity into a law is that it appears in all of the simplest, most informative true theories of the world (Lewis 1973a, pp. 72–7, see also 1986a, pp. 122–31). We can get at least presumptive evidence about what the laws are by formulating as many such theories as possible; an ideal epistemic agent provided with all of the particular facts about a world could, in principle, determine the laws from this information.

My own proposal is simple: laws of nature ought to be accepted as ontologically primitive.[3] We may use metaphors to fire the imagination: among the regularities of temporal evolution, some, such as perhaps that described by Schrödinger's equation, govern or determine or generate the evolution. But these metaphors are not offered as analyses. In fact it is relatively clear what is asserted when a functional relation is said to be a law. Laws are the patterns that nature respects; to say what is physically possible is to say what the constraint of those patterns allows.

Taking laws as primitive may appear to be simple surrender in the face of a philosophical puzzle. But every account must have primitives. The account must be judged on the clarity of the inferences that the primitives warrant and on the degree of systematization they reveal among our pre-analytic inferences. Laws are preferable in point of familiarity to such primitives as necessitation relations among universals. And if we begin by postulating that at each time and place the temporal evolution of the world is governed by certain principles our convictions about possibilities, counterfactuals, and explanations can be regimented and explained.

As an example, let us return to van Fraassen's example and variations on it. The appeal of Lewis's account is that it requires no additions to our ontology beyond particular matters of fact. Events occur in space and time, and if these events form patterns we do not swell our ontology by recognizing them. If we single out some of these regularities by somewhat pragmatic considerations,

---

[2] This view is also associated with Ramsey and with Mill. I take Lewis as exemplar because he has developed the most detailed account.

[3] To give this a conceptual level reading: the idea of a law of nature is not logically derived from, and cannot be defined in terms of, other notions.

by how they cohere to form simple and informative theories of the world, we do not add to the furniture of the world. But this account of laws fails to accord with our beliefs.

It is possible for a world governed solely by the laws that govern our world to be such that every person who reads *The Satanic Verses* is subsequently crushed by a meteor. It would take a massive coincidence, but such a coincidence will result from a possible set of initial conditions: meteors could exist on just the right trajectories. If cosmic rays were distributed in just the right way, every person who drinks milk might subsequently contract cancer. The examples could be multiplied indefinitely. The models of quantum mechanics and of Newtonian mechanics can display, due to special initial conditions or the fortuitous outcomes of random processes, accidental regularities as widespread and as striking as one pleases.

According to Lewis's view, in such a world, if the regularities are striking enough, new laws exist. The criterion for selection of laws on his view is vague, requiring a balance of simplicity and informativeness. But any such standard can be met by accidental regularities. 'All stars are members of binary star systems' is simple and highly informative, and there are models of Newton's theory of gravitation in which it is true. But for all that, we do not regard it as a law in those models: it is a consequence of special initial conditions. The Lewis view cannot admit this as a possibility, but to any astrophysicist or cosmologist it plainly is. The determination whether the universe is opened or closed will be the discovery of a simple, highly informative fact about the universe, but it will not be the discovery of a law. The initial conditions leading to a closed universe are possible, as are those leading to an open one. Whether the universe is open or closed is determined by its initial conditions and by the laws of gravitation. The fact is neither a law nor a consequence of laws taken alone.

The point is even more acute for stochastic systems. If coin flips are governed by irreducibly probabilistic laws then all sequences of results, including all heads or all tails, are physically possible. And in a world in which all of the flips happen to come up heads the proposition that they do so is both simple and highly informative. The inhabitants of such a world would doubtless take it to be a law that flipped coins come up heads, and they would be rational to do so, but they would be victims of a run of bad luck. If only finitely many coins are flipped we can even calculate the likelihood, given the probabilistic laws, of such an unfortunate event. If the number of flips is infinite the likelihood is zero, but so is that of any particular sequence. All heads is still possible in a world governed by such a law.

It is inconsistent to claim that while a law obtains the world can evolve in such a way that the law fails. If a probabilistic law governs coins then the world may evolve such that 'All coin flips yield heads' is part of the simplest, most informative theory of that world. If 'All coin flips yield heads' were a law then coins could not be governed by a probabilistic law. Hence being a member of that simplest, most informative theory cannot be sufficient for being a law.

To the ontological question of what makes a regularity into a law of nature I answer that lawhood is a primitive status. Nothing further, neither relations among universals nor role in a theory, promotes a regularity into a law. FLOTEs, along with their adjunct principles, describe how states may evolve into later states. If a law governs a particular space-time region then the physical states there will so evolve.

To the epistemological questions I must, with Armstrong, admit a degree of skepticism. There is no guarantee that the observable phenomena will lead us to correctly infer the laws of nature. We may, for example, be the unlucky inhabitants of an 'all heads' world governed by a stochastic law. We may inhabit a universe whose initial conditions give rise to regularities that we mistakenly ascribe to laws (we will see a very concrete example of this possibility in section 7). The observed correlation between distant events so important to the confirmation of quantum theory may be a run of bad luck: they could be the results of independent stochastic processes that accidentally turn out to be correlated. If so, then some of our best evidence is systematically misleading, and in rationally assessing the evidence as supporting the theory we are wandering away from the truth.

Laws are ontological primitives at least in that two worlds could differ in their laws but not in any observable respect. The 'all heads' probabilistic world looks just like another possible world that is governed by a deterministic law. If our ontology includes objective single-case propensities the two worlds will differ in their propensities. But if we wish to make laws ontologically derivative in that they supervene on the global distribution of non-nomic entities, then simply admitting objective single-case propensities will not do the job. For the probabilistic world in which these *propensities* are fixed by a deterministic law will not differ (at least in first-order propensities) from a world in which the single-case propensities are governed by a stochastic law and in which all the coin flips, beside just happening to come up heads, also just happen to get a 50 per cent propensity for coming up heads. We can introduce second-order propensities for the first-order ones, but with the obvious rejoinder. Better to regard the stochastic laws as absolutely ontologically primitive and explain single-case propensities as consequences of falling under such laws.

My analysis of laws is no analysis at all. Rather I suggest we accept laws as fundamental entities in our ontology. Or, speaking at the conceptual level, the notion of a law cannot be reduced to other more primitive notions. The only hope of justifying this approach is to show that having accepted laws as building blocks we can explain how our beliefs about laws determine our beliefs in other domains. Such results come in profusion.

The first obvious connection is to physical possibility. Our world seems to be governed by laws, at least around here. When we say that an event or situation is physically possible we mean that its occurrence is consistent with the constraints that derive from the laws. The possible worlds consistent with a set of laws are described by the models of a theory that formulates those laws.

If the laws are continuous and deterministic then the models are easily characterized. For simplicity, let us take a deterministic FLOTE and adjunct principles that operate in a special relativistic space-time. Take a surface that cuts every maximal timelike trajectory in the space-time exactly once (a Cauchy surface). Specifying a Cauchy surface is the analog to choosing a moment of time in a Newtonian regime; roughly one can think of it as a surface that cuts the space-time in two horizontally (the vertical direction being timelike). Boundary values can be specified on this surface, such as the distribution of particles, intensities of fields, etc. In some cases the data are freely specifiable, in some (due to adjunct principles) they are subject to constraints. In either case there is a well-defined class of admissible boundary values. The FLOTE now specifies how those values will evolve through time. If the FLOTE is deterministic in both the past and future directions, then the boundary values will determine a unique distribution of the physical magnitudes through all time. Such a distribution describes a physically possible world relative to those laws.[4]

If the FLOTE is stochastic then the situation is messier but still pretty clear. Specific boundary values on the Cauchy surface yield not a single model

---

[4] Lots of corners are being cut here. If the space-time is Newtonian and no constraints are put on maximum velocities, then the Cauchy surface must include a surface which surrounds a given system through all time, with 'incoming' data specified. If the laws include the General Theory of Relativity then the space-time itself is generated, not fixed. The mathematical and physics literature on boundary value problems is vast, and John Earman's (1986) is a wonderful guide to some of the intricacies, but these details do not affect our general picture. The point is that the class of models of a theory is well defined and is isomorphic to the possible worlds allowed by the laws described by the theory. Further, data on a small portion of a world together with the laws can determine data throughout the world, or throughout a large region.

but a set of models, corresponding to all of the outcomes permitted by the laws. Furthermore, the set of models consistent with some boundary values is invested with a metric over measurable subsets, a measure of how likely it is, given those boundary values, that the world will evolve into one of the members of that subset. If the boundary conditions specify that 100 coins are about to be flipped then the set of models associated with the probabilistic law of unbiased coins (50 per cent likelihood of heads) and with a law of biased coins (e.g. 80 per cent likelihood of heads) are identical: they each contain a model for each possible combination of heads and tails. But the probabilities assigned to these models by the two theories differ.[5]

Given a FLOTE and adjunct principles, then, the notion of physical possibility relative to those laws can be immediately defined. And the result is intuitively correct. Newtonian mechanics allows for a world in which all spheres are gold but it is not a law that all spheres are gold. Stochastic laws allow for the possibility of uniform results that are not a matter of law. The objections to the Lewis view are avoided and the intuitions that backed the objections are explained.

Above it was suggested that the notion of a stochastic law should be taken as more primitive than that of an objective single-case propensity. Propensities can now be easily derived. The propensity for some occurrence, given a set of boundary conditions on a Cauchy surface, is just the probability assigned by the stochastic laws to the set of models with those boundary conditions and that outcome. Given some actual model, it may be the case that the propensity assigned to an occurrence relative to any Cauchy surface that cuts within some temporal distance D before the occurrence is the same, or that the propensities assigned by surfaces that cut within D approach a limit as D approaches zero. If so, then that limit is the objective single-case propensity for the event.

As a small bonus, we can also see what is so peculiar about the distant correlations of quantum mechanics. Take a model of a stochastic theory defined on a relativistic space-time. Consider an event *e*. Take any two Cauchy surfaces that overlap where they intersect the back light-cone of *e*, although they may differ outside the back light-cone. We will say that *e* is the result of

---

[5] The official definition assigns the probability to subsets of the set of models rather than to individual models because for stochastic processes with a continuous range of possible outcomes (e.g. radioactive decay, which can occur at any time), or models with an infinite number of discrete stochastic events, the probability of each model may be zero. Still, specified subsets, such as those in which the atom decays within a particular time period or those in which a particular discrete event has a given outcome, may be assigned a definite probability. Of course, individual models may be assigned non-zero probabilities by certain theories.

a *local stochastic process* if the probability assigned to e by the theory relative to any such pair of Cauchy surfaces is the same. The notion of a local stochastic process is slightly weaker than that of having an objective propensity. If we add that the probabilities assigned by all continuous sequences of Cauchy surfaces should approach the same value as the maximum timelike interval between the event and the intersection of the Cauchy surfaces with the back light-cone approaches zero, then any result of a local stochastic process will have an objective probability. That is, in such a case not only will variations in a Cauchy surface outside the back light-cone not affect the probabilities, also how the surface cuts across the back light-cone does not affect them.

The motivation of the definition is straightforward: according to (at least one interpretation of) Relativity, events should be influenced only by other events in their back light-cone. So Cauchy surfaces that agree on the back light-cone and have the same boundary values there should agree on everything that could possibly be relevant to the occurrence of the event. Therefore the theory should assign the same probability to the event on the basis of data on any such Cauchy surface.

The problem with quantum mechanics is now quickly stated: the probabilities for correlations between distant events assigned by quantum mechanics cannot be reproduced by any theory in which all events are the result of local stochastic processes. Quantum mechanics cannot be reconciled with the interpretation of Relativity stated above.

FLOTEs give us immediate purchase on the notions of physical possibility and objective propensity. And once we have the set of models, physical necessity is easily defined in the usual way, using the models as mutually accessible possible worlds. So in a way, the problematic inherited from Hume has been solved by turning the direction of analysis around. Here is what I mean.

Hume begins by investigating the notion of cause and effect, and finds within it a notion of necessary connection between events. He then worries about giving an empiricist account of the origin (and hence content!) of the notion of this necessary connection, and finds that he is led either to constant conjunction between events or to a subjective sensation that accompanies an inference bred of long experience of constant conjunction. The 'necessity' must reduce either to mere pattern or to a purely subjective sensation, and in neither case pertains solely to the two events thought to be necessarily conjoined. Although Hume does not focus so intently on the motion of a law of nature, the natural implication is that laws can be nothing but patterns of events either.

A Modest Proposal

I take content of the laws to be expressed by equations like Newton's equations of motion, and the status of lawhood to be primitive. What then of the notion of 'necessary connection'? The content of the laws can be expressed without modal notions, and suffices to determine a class of models. The models can then be treated as 'possible worlds' in the usual way, and so provide truth conditions for claims about nomic possibility and necessity. The laws themselves, of course, turn out to be nomically necessary, since they obtain in all the models. We can give a clear account of the 'had to' in claims like 'The bowling ball had to fall once the support beneath it was removed': in every model of the laws with an unsupported bowling ball, the bowling ball falls. So we have all the 'physical necessity' we need without having invoked anything beside the laws. And given the laws we can easily construct truth conditions for even more.

5. FLOTES AND COUNTERFACTUALS

One who regards laws as primitive will also regard them as quite definite. At a given time the future temporal behavior of the world is constrained in some exact way by the laws, irrespective of our beliefs, desires, and concerns, irrespective of pragmatic considerations or contexts. On the other hand, the evaluation of counterfactual claims is widely recognized as being influenced by context and interest. This contrast poses a challenge for those who seek to explicate the bearing of laws and counterfactuals on one another. The challenge is more difficult for anyone who holds that our judgements about laws depend derivatively on previously settled beliefs about counterfactuals: it is not easy to create a rigid structure if some of the basic components are elastic. In the opposite direction thing go more smoothly, for if judgements about counterfactuals depend on beliefs about laws and on other things, and if the other things reflect pragmatic considerations, then judgements about the counterfactuals may be variable or indefinite although the beliefs about laws remain fixed.

What follows is not a unified theory of all counterfactuals. It is likely that no such theory exists. What would have happened if it had been 20° warmer at Canaveral on the day the *Challenger* exploded? If trees could walk? If Plato and Carnap had been contemporaries? If gravity were an inverse cube force? If one could trisect the angle with ruler and compass? We are wont to consider what would have happened if physical conditions had been different, if laws were changed, if metaphysical necessities were violated, even if logical truths failed to hold. The methods of evaluation of these claims are bound to be diverse. Even Lewis's theory of counterfactuals must give out in the last case: it is non-trivially true that if one could trisect the angle one could square the circle. Lewis can get truth, but not non-triviality: his theory cannot explain why a mathematician would try to prove the claim (Lewis 1973a, pp. 24, 25). And Lewis can certainly not explain discriminations of such counterfactuals as true *or* false, e.g. it may be false that if one could construct a 23-gon one could square the circle. It is bootless to rest the semantics of counterfactuals on relations to possible worlds in this case.

Still, a large number of types of counterfactuals seem to be treated similarly, and, more important, those so treated are among those for which our intuitions are strongest. We will begin with cases that are as uncontentious as possible and are closely tied to physical law. If we understand what makes these cases uncontroversial we will be able to predict under what circumstances doubts about the truth value or the meaningfulness of counterfactuals will begin to creep in.

If the bomb dropped on Hiroshima had contained titanium instead of uranium it would not have exploded. If Lindbergh had set out with half as much fuel, he would not have made it across the Atlantic. If the ozone layer in the atmosphere should be destroyed, the incidence of cancer will increase. These claims seem undoubtedly true. In the agora of everyday discourse and in scientific contexts such claims are treated on a par with descriptions of the *Spirit of St Louis* and assessments of the chemical composition of the atmosphere. Even if they are ultimately shown to be counterfeit currency, passing as statements of fact when they play some other role, one must account for the assurance with which such claims are made and evaluated. (In defense of their legitimacy it is notable that the last, the future subjunctive, may or may not be counterfactual for all we know now. That doesn't affect the means we use to verify it.)

I take the three sentences above to be true, and I take their truth to depend on the laws of nature. They also depend on other factors. Let us start with the first case.

We have already seen how FLOTEs plus boundary conditions on a Cauchy surface generate models. For simplicity, suppose that the laws governing our world are local and deterministic, as they essentially are for the case at hand. Extension to stochastic laws will come later.

We wish to know what would have happened if the bomb had contained titanium in place of uranium. Here is the recipe. Step 1: choose a Cauchy surface that cuts through the actual world and that intersects the bomb about the time it was released from the plane. All physical magnitudes take some value on this surface. Step 2: construct a Cauchy surface just like the one in Step 1 save that the physical magnitudes are changed in this way: uranium is replaced with titanium in the bomb. Step 3: allow the laws to operate on this Cauchy surface with the new boundary values generating a new model. In that model, the bomb does not explode. Ergo (if we have got the laws right, etc.) the counterfactual is true.

In this three-step process laws come into play essentially, but only at the last stage. If we manage to select a unique Cauchy surface and to alter its data in a unique way, and if the laws are deterministic, then all counterfactuals with that antecedent will have determinate truth values. For a single model will thereby be specified and the consequent of the counterfactual will either be true in it or not.

Even if there is some ambiguity in the Cauchy surface and in the way to change the boundary values, still the claim may have a determinate truth value. Because of the ambiguity many different surfaces may be selected and the data on them changed in many ways. The result is a set of models rather than one: a model for each acceptable changed surface. If the consequent obtains in all of the models, the counterfactual is true; if in none, false. If it obtains in some but not others, the counterfactual has an indeterminate truth value and our intuitions should start to get fuzzy. Examples will bear these predictions out.

The purpose of the antecedent of a counterfactual is to provide instructions on how to locate a Cauchy surface (how to pick a moment in time) and how to generate an altered description of the physical state at that moment. The antecedent is not an indicative sentence at all; it is more like a command. If the command is carried out a depiction of a situation will result and according to that depiction a certain indicative sentence will be true, but the command is not the same as the indicative sentence. Despite surface similarities to the material conditional, the counterfactual conditional is not a two-place function whose arguments are two propositions. It is a function whose first argument is a command and second is a proposition. This will explain why the counterfactual conditional fails to have any of the formal features of the material conditional: transitivity, contraposition, and strengthening of the antecedent (cf. Lewis 1973a, pp. 31 ff.).

Thus the 'if the bomb dropped on Hiroshima had contained titanium instead of uranium ...' directs us to a moment shortly before the atomic explosion and instructs us to alter the description of the state of the world so that titanium replaces uranium in the bomb. And there is a tacit ceteris paribus condition: leave everything else the same. Don't fool with the position of the plane or the wiring of the bomb or the monetary system of France or anything else. Similarly, if I command you to take the roast beef in to the guests you have not carried out the command if you step on the roast beef first, and if you murder one of the guests in the process you did not do so on my instructions. Of course, one cannot just change the uranium into titanium and leave *everything* else the same. The total amount of titanium in the universe will be changed, as will the ratio of titanium to steel in the bomb. So the clarity of the instruction, what counts as fulfilling it and what not, depends on the clarity of the ceteris paribus clause, and this is a function of the context.

The ideal case obtains when the physical state of the world is understood as specified by providing values of a set of independent variables and the command is to change one of those variables. Then 'ceteris paribus' means 'leave the rest of the variables alone'. In Newtonian mechanics, such variables could be the positions, masses, and velocities of all the particles. An instruction to change the mass or velocity of a certain particle will be, in this context, unambiguous. But such an analysis of the physical state into a set of independent variables is not provided by the laws or by Newtonian theory as a whole. We can specify states by giving masses, positions, and velocities, or equally well by masses, positions, and momenta. If I am told to increase the mass of a particle this difference may have an effect: increasing the mass while leaving the velocity unchanged increases the momentum; increasing the mass while leaving the momentum unchanged decreases the velocity. In each case different *cetera* are *paria*, and which change is appropriate is decided, if at all, by context and background assumptions.

The independent variables might, in principle, be quite bizarre. Instead of independently specifying the masses of each particle, we could give the mass of a chosen particle and the mass ratios of the rest to the standard. In this context, increasing the mass of the standard particle and leaving everything else the same entails increasing the masses of all the others. Of course, a counterfactual instruction would have to warn us if this were meant, for the choice of variables is highly unusual.

In most contexts we have a rough and ready sense of which variables are to be treated as independent of one another. But the instructions may be vague in other ways. 'If Lindbergh had set out with half as much fuel…' tells us to choose a moment near Lindbergh's departure and change the state of the world so his fuel tank contains half as much gasoline. Don't change his hair color or the structural members of the *Spirit of St Louis* or the height of the Eiffel tower. But we are to delete the extra fuel and replace it with ... what? Not a perfect vacuum. Not water. Air, presumably. But the instruction doesn't tell us to replace it with air, so this information must come from background assumptions. When such background assumptions do not suffice to clarify the means of carrying out the command, the counterfactual may fail to be determinate.

Just as the instructions for Step 2 may be more or less vague, so may the instructions for Step 1, choosing the moment of time. Commonly this ambiguity will make no difference to the outcome, sometimes it will. Consider some data. What are the truth values of the following counterfactuals? Case 1. Tom is standing in a field two meters north of Sue. A large meteor crashes down, incinerating Tom. Counterfactual A: If Sue had been standing two meters north, she would have been killed. Case 2. Tom, a famous politician, is standing two meters north of Sue. A crack sniper, who has been stalking Tom, fires and assassinates him. Counterfactual B: If Sue had been standing two meters to the north, she would have been killed. Counterfactual C: If the trajectory of the bullet had taken it two meters south, Sue would have been killed. Counterfactual D: If the bullet had been deflected two meters south, Sue would have been killed.

The reader is requested to consult her or his intuitions on these cases. The author and several colleagues questioned are in broad agreement. Counterfactual A is clearly true. Counterfactual B is hazy, a common comment being 'needs more context'. Counterfactual C is probably true, although this is not so clear as case A. D is clearly true.

The recipe delivers these results. Return to any moment shortly before the meteor strikes and move Sue two meters north, in any reasonable way. Dump Tom somewhere else. The trajectory of the meteor is unchanged and the laws generate a situation in which Sue is hit. In the second case, though, the exact moment chosen is of great importance and different choices give different outcomes. If a moment is selected after the shot is fired but before it hits Tom, Sue will be shot. If the moment is somewhat before the shot was fired, the natural laws yield a different outcome. The sniper will notice that he is not aiming at Tom and will compensate. Sue will be safe. The third case has both the ambiguity of time and of the instruction. How should we alter the trajectory? If by altering the aim of the gun then we have two cases. We can keep Sue's position unchanged, and she will be hit. Or we can keep the expertise of the marksman unchanged, in which case he would only be aiming where Tom is, so Tom must be moved. The expertise of the sniper is especially important because it is one of the few feature of the situation explicitly stated. So C can get different outcomes. D is entirely clear: the bullet is to be changed en route, so the expertise of the marksman is not in question. All models yield the same result.

An instruction or command is not a proposition, nor can it be construed as an attitude adopted toward a proposition. At least this is so if one considers 'A weighs the same as B', 'B weighs the same as A', and 'A and B weigh the same' to express the same proposition. For instructions command not only that one change a state so as to make a certain proposition true, they also indicate how the change is to be effected. Roughly, 'If S were P ...' instructs us to change S so that it is P. Examples: (1) If Laurel weighed the same as Hardy, the pair would have weighed over 500 pounds. (2) If Hardy had weighed the same as Laurel, the pair would have weighed under 400 pounds. (3) If Laurel and Hardy had weighed the same ...? (In these cases, evaluating the counterfactual requires no third step since the truth value of the consequent is already determined at Step 2.) Instruction 1 tells us to change Laurel to make the proposition true, instruction 2 to change Hardy. Instruction 3 is unclear. To evaluate counterfactual 2 by fattening Laurel up at Step 2 would be as perverse, and as incorrect, as a weight expert who set about fulfilling the command to make Hardy weigh the same as Laurel by feeding Laurel.[6]

Different contexts can demand that counterfactuals be evaluated by methods that accord to a greater or lesser degree with the recipe. Those more in accord with it give the more correct counterfactuals about what would actually happen. Examples: (a) The ant can lift 500 times its body weight. So an ant the size of a human could lift a VW bus. (b) Weight increases as the cube of linear dimension. Tensile and compressive strength of structural materials increases as their cross-section, i.e. as the square of linear dimension. Hence an ant the size of a human would collapse under its own weight (see Galilei 1974, second day). The first counterfactual stipulates (by context) the factor to remain unchanged at Step 2: weight-to-lift ratio. The consequence follows by logic after Step 2, Step 3 is unneeded. The second scales up the creature at Step 2 and then lets the laws governing the materials determine the outcome. It is the correct result if we are concerned with physical possibility. A small model of a skyscraper made of paper may stand up perfectly well.

---

[6] This problem, as well as most of the others discussed here, was pointed out by Nelson Goodman (Goodman 1983, chapter 1).

A scaled-up full-size model made of scaled-up paper would collapse. That would also be the fate of a gargantuan ant.

We can now explain the failure of counterfactuals to display the formal features of the material conditional. Failure under strengthening of the antecedent is manifest. If Lindbergh had had half as much fuel he would not have made it. If he had had half as much fuel and an engine twice as efficient he would have. The first instructs us to reduce the fuel and leave the engine alone, the second to change both fuel and engine. It is hardly surprising that the results of each instruction might differ radically.

Failure of transitivity is equally manifest. If A had occurred, B would have occurred. But the *way* B would have occurred might be very different than the way we would bring it about if instructed to. If the Earth had exploded in 1987, Ivana would not have found out about Marla. If Ivana had not found out about Marla, Trump would be a happier man now. In normal contexts both of these are true. It does not follow that if the Earth had exploded, Trump would be happier. The instruction to go back and prevent Ivana from finding out is vague in the extreme; one could bring it about by innumerable different mechanisms, but apocalyptic cataclysm is not among them.

There is a context in which transitivity holds, which is when we are *continuing* a scenario. If A had occurred, B would, and if B had occurred *in that way* C would. So if A had occurred C would. Because of this, we may feel a bit queasy about evaluating the second counterfactual above given its proximity to the first, even though in neutral contexts we would readily assent to it.

Contraposition fails for the same reasons. If the antecedent and consequent describe events at different times then one of the counterfactuals must take the more unusual form 'If A had happened B would have had to have happened.' Consider an undecided voter who decided at the last moment, and with little conviction, to vote for Reagan. Several years later, unhappy with the state of the country, she consoles herself: 'Well, even if I hadn't voted for him he still would have been elected.' This is true. The contraposition, 'If Reagan had not been elected, then she would have had to have voted for him,' is certainly not true, and one could make a case that it is false.

It is also obvious that contraposition must fail on purely syntactic grounds. The consequent of a counterfactual is a proposition, so the truth value of a counterfactual is not changed under substitution in the consequent of sentences that express the same proposition. We have already seen that this is untrue for the antecedent. Under contraposition, the sentences in the antecedent and consequent change roles, so changes that can make no difference for the original counterfactual can change the truth value of the contraposition. If contraposition always held, we could counterpose 'If Laurel weighed the same as Hardy' in the antecedent to 'Laurel would not have weighed the same as Hardy' in the consequent, switch the consequent to 'Hardy would not have weighed the same as Laurel', then counterpose this back to 'If Hardy weighed the same as Laurel' in the antecedent, all without changing the truth value. But we have seen that this is impossible.

So far we have considered only cases where all of the ambiguities and vaguenesses result from the instructions on how to carry out Steps 1 and 2 of the recipe. At Step 2 we have a set of Cauchy surfaces with changed boundary values, which the laws of Step 3 extend forward (or backward) in time to yield a model. Rather deep difficulties appear when we attempt to extend this account to stochastic laws. These difficulties are accompanied by a divergence of opinion about the evaluation of such counterfactuals, a divergence of opinion that can now be explained.

At first glance the effect of stochastic laws seems to be the same as that of ambiguity about which Cauchy surface to choose and vagueness in the instructions for changing the boundary data. In these cases the result is to generate a set of models instead of one, and the counterfactual only has a determinate truth value if the consequent obtains or fails to obtain in all of them. So too, if we precisely specify a Cauchy surface and the way data are to be changed, still stochastic laws will generate a set of models rather than one. Certainly, if the consequent obtains in all the models then the counterfactual is true, false if it obtains in none. And one can also maintain that the counterfactual has no determinate truth value if the consequent is true in some models and false in others. Michael Redhead, for example, takes just this view when discussing counterfactuals and quantum phenomena (Redhead 1987, pp. 90 ff.). As Redhead points out, this has a result which some consider counterintuitive. Suppose the world is governed by completely deterministic laws save for one that is stochastic: flipping a coin gives a 50 per cent probability of heads, 50 per cent tails. This probability is unaffected by all attendant circumstances. Further suppose that a flipped coin actually lands heads. What of the counterfactual: had I raised my arm a moment before the flip, it still would have come up heads?

On Redhead's analysis the counterfactual is not true, even though raising the hand may be acknowledged to have no causal influence on the coin. Following the recipe, we go back to a moment before the flip and change the data so that the hand is up rather than down. Letting the laws operate on the new data, we get two models, one with the coin heads the other with it tails. So application of the recipe yields results that have been endorsed by at least some philosophers.

The result is not entirely happy, though. It causes deep difficulties for any counterfactual analysis of causation. And it conflicts with what David Mermin has called the Strong Baseball Principle: 'The strong Baseball Principle insists that the outcome of any particular game doesn't depend on what I do with my television set—that whatever it is that happens *tonight* in Shea stadium will happen in exactly the same way, whether or not I am watching it on TV.' (Mermin 1990, p. 100). The Strong Baseball Principle is supposed to apply even if tonight's game involves some irreducibly stochastic events.

Applying the Strong Baseball Principle to the coin flip, whether or not I had watched the flip, it would have come out the same. Since raising the arm has by hypothesis no influence at all on the flip, we get the result that even had my arm been up, the coin still would have fallen heads.

Redhead's analysis, unless amended, conflicts not only with the Strong Baseball Principle but with the requirement that a subjunctive conditional whose antecedent turns out to be *true* reduce to a material conditional. Should my arm have been down, the coin would have come up heads—after all, my arm *was* down and the coin *did* come up heads. But if we follow the recipe, we don't get this result. We choose a Cauchy surface and find at Step 2 that the command has already been carried out and no changes need to be made. Allowing the laws to operate, we again get two models, one heads, one tails.

In deterministic cases, all occurrences are fixed by the boundary values and the laws. The tacit ceteris paribus condition applies only at Step 2: we are to effect the commanded change making as few collateral changes as possible. In stochastic cases, events are determined by boundary values, laws, and the random outcomes of stochastic processes. How are we to apply the ceteris paribus clause to this last element?

It is too strong to suggest that we should restrict our attention to models in which as many events as possible match those in the actual world. There is no determinate fact about how the coin would have fallen had we done something to it, such as flipping it higher or starting the flip in a different position. If there had been a different causal process leading to the result we might have gotten a different result. If the process is unchanged in the counterfactual, as in the case of raising my hand, so should the result be. The question is how we can determine which causal processes would be changed by a change in the boundary data.

A solution to this problem can only be sketched here, and perhaps never can be resolved by philosophical means. I hesitate to call what follows a theory; it is rather a description of how we may think about these cases. It explains why we might subscribe to the Strong Baseball Principle and it yields the result that if nothing had been changed, if a subjunctive conditional has a true antecedent, then the coin would be heads.

If I had raised my hand, the only results of stochastic processes that might have had different results are those that would have been (or might have been) affected by my hand going up. Any stochastic process is an evolving set of physical magnitudes. Let us call any physical magnitude which is unchanged when we apply Step 2 *uninfected*; those which are changed are, in both the original and new data sets, *infected*. This distinction may be quite clear in some theories, in others (notably in certain quantum states) the physical magnitude may not be localized and may be so entangled that the infection cannot be localized. In such cases, our intuitions break down.

The laws tell us how later magnitudes evolve from earlier ones. In some cases (again not in all, and perhaps not in quantum mechanics) a later magnitude can be seen as being generated from a set of earlier ones via the laws. Any magnitude generated at least in part from an infected magnitude is also infected. On one interpretation of the no-superluminal-signals constraint associated with Relativity, Relativity requires that all local magnitudes be generated from magnitudes exclusively in their back light-cones. In a pure particle theory with local interactions, infection could only be spread by contact with infected particles. So in at least some cases the generation of later states from earlier ones is of the right form to trace infection through time. In the actual world some of the later magnitudes will be infected, and in each of the models that results from the application of Step 3 the infection will spread in some way. Our ceteris paribus clause can now be stated: starting from the new Cauchy data and going forward in time, whenever the models that result from Step 3 start to diverge, if the divergence is due to the outcome of processes which, until that time, are identical and are uninfected in all of the models and in the actual world, then we should keep only those models in which the outcome agrees with the outcome in the actual world. If in any model the process has become infected, we must keep all the models generated by that process.

This prescription gives some good results. If the antecedent to a subjunctive is true, then Step 2 leaves all magnitudes uninfected, and we get back a single model that matches the actual world. If the magnitudes that are changed by watching rather than not watching TV do not propagate to Shea stadium, then the result of the game would have been the same no matter what I did. And if we deny this, saying that the magnitudes do propagate, then my intuition dissolves: perhaps I might have made a difference.

Although the infection theory sounds a bit complicated, it fits well one way of reasoning about counterfactuals. We start with the original situation and *update* it to fit the counterfactual condition. The first step in the update is Step 2 of the recipe. Then we consider what collateral changes those changes would have brought about. Our beliefs about the laws of temporal evolution guide this process. For an event that is remote from the region of the change, only a law connecting the events changed to that distant event would provoke us to update the model there. This is true even if the laws governing the situation are indeterministic. The difference between Redhead's and Mermin's intuitions is exactly the difference between those who *generate* a new model after Step 2 and those who *update* the original model on the basis of Step 2.[7]

The three-step recipe can also explain another sort of counterfactual: the counterlegal in which we assert what would happen were the laws to change. Step 2, rather than changing boundary values, alters instead the laws. Again a ceteris paribus condition accompanies the command: if told to change the gravitational constant we should leave the laws of electromagnetism alone. From the unchanged boundary conditions and the new laws we apply Step 3. The algorithm gives us the truth of 'If the gravitational force should disappear, the Earth would not continue to orbit the Sun.' All the problems of vagueness of the instructions reappear here.

Finally, we should note that although many counterfactuals involve consequents that occur later than the antecedent, some do not. If the bomb dropped on Hiroshima had contained titanium rather than uranium then a bomb containing titanium would have had to have been put on the plane. This can be accommodated in several ways. On the one hand, laws can in principle sometimes be used to retrodict as well as predict. In this case, one could use the recipe in unchanged form: change the Cauchy data and evolved backward rather than forward in time. But our intuitive sense of the world relies on LOTEs that typically work only in one direction. For example, given a glass of water with ice in it, on a hot day, we easily predict that if left alone the ice will melt and the water become lukewarm. But given a glass of lukewarm water on a hot day, and assured that it has been left alone over

---

[7] The distinction between updating and constructing does not make a difference if the laws are deterministic since only one model is consistent with the laws and the new boundary values.

the past hours, we would not hazard a guess about whether a few hours ago it contained ice. So whether or not the fundamental laws of physics are time reversible, the phenomenal laws we are familiar with are not. The natural tendency is not to evaluate counterfactuals whose antecedents postdate their consequents by 'running the laws backwards'—we wouldn't know how to do that—but rather by searching for the sorts of earlier states which would, in accord with the forward-running laws, lead to the later state. The only obvious way to *get* the titanium bomb onto the flying Enola Gay is to have had it *put* there earlier: one runs many scenarios forward in time rather than trying to run a single scenario (with the bomb on the plane) backward.

The vastness of the literature on counterfactuals does not allow an even comically inadequate comparison of this theory of (certain) counterfactuals with all the other proposals that have been made. I would, however, like to point out some contrasts with David Lewis's theory. Lewis's analysis postulates a context-dependent metric of overall similarity between worlds. Counterfactuals are evaluated by going to the nearest worlds in which the antecedent holds, and seeing if the consequent holds there. The theory presented here obviously has its roots in this approach. But Lewis's theory has been widely criticized on the grounds that no unprejudiced judgements of overall similarity yield the right result.

In deterministic contexts the problem is bad enough. Lewis concedes that in tracing out the consequences of a counterfactual condition in a deterministic world, the laws must be respected. In most details of particular fact, a world in which a titanium bomb is dropped *and explodes just like the actual uranium bomb* is much more similar to the actual world than one where it is dropped and doesn't explode. This is the wrong sense of similarity. So keeping the laws the same must take priority in judging the temporal evolution of worlds. On the other hand, what we call the changes due to Step 2 often require violations of law. We deflect the bullet without assigning a physical cause. Lewis wants these changes, so he admits that the nearest worlds will contain miracles, that worlds containing some miracles are closer to the actual world than ones that are thoroughly law governed. So what has become of the priority of law? Apparently, in the similarity metric some miracles are more equal than others.

Lewis employs some ingenious casuistry of miracle-comparison in an attempt to get the right result: some miracles are bigger or more widespread or more noticeable than others (Lewis 1986a, pp. 38 ff.). But none of those qualitative distinctions gets the right result: the miracles we don't care about are the ones *we* bring about in carrying out Step 2. *We* deflect the bullet because we are instructed to. Once we have the new data on the Cauchy surface (or on the pair of surfaces, one below and the other above the region which we change) then all that matters is the laws. And the changes that take place on that surface, or in the region between the pair, are justified by reference to the instructions, not by appeal to similarity. Indeed our recipe makes no reference to an overall similarity between worlds, the nearest thing being a ceteris paribus condition that determines what counts as the appropriate carrying out of a command.

If the problems in deterministic worlds are bad, those in stochastic worlds are insurmountable. It is not certainly true that had we started the coin flip in another position it still would have come up heads. But the model in which it does come up heads must be closer than the one in which it doesn't according to any similarity metric. For the two models match in point of lawfulness: the change in the boundary conditions is the same in both and each is an equally lawful, indeed equally probable, continuation of the boundary conditions. But the heads world is more similar to the actual world in point of particular fact, and if other events are decided deterministically on the basis of the flip, the differences in the two models may be massive. Lewis's analysis has no choice but to brand 'If we had started the coin in another position it would still have come up heads' true, but it clearly isn't. Our resolution of these cases by reference to infection makes no use of any comparisons of overall similarity.

Finally, we should note that Lewis's theory is concerned only with the semantics of counterfactuals, with the metaphysical conditions which make counterfactual claims true or false. He does not directly concern himself with the psychological question of how people evaluate counterfactuals, what processes underlie their intuitions. It seems quite unlikely that the psychological process could mirror Lewis's semantics: people do not imagine a huge multiplicity of worlds and subsequently judge their similarity and pick out the most similar. Rather we *construct* representations of possible worlds from the counterfactual description, construct them according to the three-step recipe or something like it. We imagine the proposed situation and let the scenario unfold in our mind, guided by the principles we accept as governing temporal evolution. Since the principal test of a semantic theory is how it accords with our intuitions, a semantics modeled on the process that generates intuitions is likely to be more satisfactory than one that ignores psychological procedure.

If counterfactuals are judged by a process resembling the three-step recipe then it is clear why, on the conceptual level, laws support counterfactuals.

We construct the counterfactual situation by means of the laws, so the laws must hold. On the ontological level, if the semantic values of counterfactuals are fixed by facts about laws in a way described by the recipe then the bearing of laws on the truth of counterfactuals is manifest.

6. MODELS, LAWS, AND EXPLANATIONS

There are two other views concerning laws and scientific theories that deserve our attention. They are in many ways opposed to one another, yet share similarities which permit them to be grouped together fruitfully for our purposes. The first is an account of laws championed by Storrs McCall and Peter Vallentyne.[8] The second is Bas van Fraassen's recently defended view that laws can be dispensed with altogether.

Vallentyne's theory begins with the notion of a world-history and an initial world-history. A world-history is 'a state of affairs that involves everything that happens at all points of time in some world'. An initial world-history is 'a state of affairs that involves everything that happens up to some point of time in some world, and involves nothing pertaining to later times. World-histories and initial world-histories involve only what happens; in particular, they involve nothing concerning nomic features of the worlds' (Vallentyne 1988, p. 604). A nomic structure is a relation between world-histories and initial world-histories. Roughly, it is the relation that holds between an initial world-history and a world-history iff the world-history is a continuation of the initial world-history that is permitted by the laws that govern the world. A law is any statement guaranteed to be true by the nomic structure.

Much could be said about the Vallentyne/McCall project, but the central issue is contained in the 'iff' sentence in the paragraph above. For Vallentyne it is an explication of the right-hand side by the left, whereas I have presented it as an explication of the left-hand side by the right. If we believe in FLOTEs (and adjunct principles) then we believe in models of FLOTEs (world-histories) and in a relation between models of a FLOTE and their initial segments. On this account, laws are primitive and nomic structures can be defined from and explained in terms of them. On Vallentyne's account it is the other way round.

---

[8] As with the Mill–Ramsey–Lewis theory and the Armstrong–Dretske–Tooley theory, I will use the writings of one representative, in this case Vallentyne 1988.

I certainly cannot criticize any account for having primitives that are not further explicated. But I do think one can pass judgement on the intuitive clarity of the primitives. Anyone who has studied physics or biology or chemistry or economics has a notion of what a law is. Anyone who has watched plants grow understands what it is to take 'plants grow toward the light source' as a non-accidental fact about plants, a feature of their development that allows us to predict what a plant will do and to judge what it would have done. It is from individual facts like these that we build up a picture of how the world might go, what the world-histories might be, not vice versa. It is very hard to see what a *primitive* grasp of a nomic structure might be, a grasp that might be obtained before the notion of law is in play.

It is also highly dubious that sense can be made of a world-history antecedent to the notion of a law. Laws and the physical magnitudes they govern are to some extent mutually determining. It makes little sense to suppose that an electromagnetic field exists in a region of space if the object in that region does not at least approximately obey Maxwell's laws, does not deflect charged particles, etc.[9] So if world-histories involve claims about electromagnetic fields and charged particles, then the laws must already be settled to a large extent by the world-history before any question about a further relation between the world-history and its initial segments can be raised. When Vallentyne says that world-histories 'involve nothing concerning the nomic features of the worlds' it is unclear what sorts of states of affairs, if any, can meet this requirement.

Finally there is the question of strength. We have already seen that an account that takes FLOTEs as primitive can do all of the work of Vallentyne's view since the nomic structure can be defined from the FLOTEs. Can we inversely define the laws, and particularly the FLOTEs, given only the nomic structure?

Laws are supposed to be propositions whose truth is guaranteed by the nomic structure, propositions true in all the world-histories with a given nomic structure. The FLOTEs should certainly turn out to be laws (I don't know how to prove this except by generating the world-histories from the FLOTEs because I don't know how else to identify the relation in question). But are all the other propositions whose truth is guaranteed by the nomic structure laws?

They are not if one of the classic criteria for laws is accepted: confirmation by positive instances. By Vallentyne's criterion, 'Everything which is either a

---

[9] In tribute to Wilfrid Sellars's parallel claim about concepts, we might dub this doctrine 'Physical Magnitudes as Involving Laws and Metaphysically Impossible without Them'.

cow or a sample of Uranium 238 is either a mammal or radioactive' is a law. But observations of any number of one sort of positive instance (cows) gives us no reason to believe anything about others (uranium). Confirmation is not transmitted to all unobserved instances. It is natural to regard the claim above as a consequence of two laws which is not itself a law. Vallentyne's definition does not allow this.

More controversially, I would object to calling certain propositions laws even though they are confirmed by their positive instances. 'All humans who live in houses with prime house numbers are mortal' is not a law because the class referred to is not a natural kind. Positive instances do transmit confirmation to unobserved cases, but only because they confirm 'All humans are mortal,' and all instances of the first claim are instances of the second. The proposition is not a law but again a consequence of a law, this time by conjunction rather than disjunction.

Even more controversially, I don't think that 'All humans are mortal' is a law of nature even though it is couched in terms of natural kinds and properties, confirmable by positive instances, and guaranteed by the nomic structure. It too is a consequence of laws, laws of biochemistry, physiology, etc., but is not itself a law.

Being purely cranky, one could challenge the advocate of the lawhood of 'All humans are mortal' to find the generalization treated as a law, or even stated, in a biology text. One might find it remarked that humans are mammals, omnivores, etc., but not that they are uniformly destined to die. The equally cranky response would be that of course one never writes that all humans are mortal because it *goes without saying* that all humans are mortal. Everyone knows it. So why waste ink printing it? But the supposed law is not missing just because it is tacit. No biological fact is ever *explained* by reference to this law, whether the reference be tacit or explicit. In particular, no one's *death* is ever explained by reference to this law. People die of cancer, or stroke, or trauma, or asphyxiation. Nobody ever dies of humanity.

At the beginning of this article I remarked that one of the widely acknowledged conceptual connections is between laws and explanations. The covering law model of explanation took this connection as its primary inspiration. We need not review the shortcomings of this model: not every subsumption under law, not every derivation of a fact from laws and initial conditions, constitutes an explanation. Nonetheless, there are distinctive explanations whose primary structure consists in showing how the operation of laws of nature led to a particular event or state of affairs. We understand why the planets obey Kepler's laws (as well as they do) when we derive the trajectories of bodies with certain initial velocities and masses from the General Relativistic laws. The operation of a FLOTE explains why certain physical magnitudes take on values at later times given their values at earlier times. Not every subsumption is an explanation, but a general proposition that cannot be used to explain any of the instances it covers can hardly be called a law.

Of course, given the right context the statement 'All humans are mortal' could play a role in providing understanding of a fact. But given the right circumstances so could 'All the coins in N.G.'s pocket on VE Day were silver.' The issue is whether the explanatory power derives merely from the truth of the assertion or from its status as a law. The explanation of individual cases by subsumption does depend on lawhood, for it is acknowledged that subsumption by accidental generalizations does not explain. In the 'all heads' world governed by stochastic laws it in no way explains why a particular coin came up heads to note that they all did. Indeed, since the laws are by assumption irreducibly stochastic, there *is* no explanation of why a particular coin came up heads. In the deterministic 'all heads' world, the particular events are explained by the law. We might, of course, seek a further explanation for the law.

This picture of explanation by nomic subsumption faces a severe challenge in Bas van Fraassen's *Laws and Symmetry* (1989). Van Fraassen urges us to abandon all philosophical accounts of laws as unworkable, and to abandon laws altogether, seeking an account of scientific practice and of theories that eschews all talk of laws. Much of van Fraassen's inspiration derives from the observation that many laws (especially conservation laws) that had been postulated as primitive in classical physics are now derived from underlying symmetries. This aspect of van Fraassen's view will not come in for further notice here.[10] What are of interest are his views on scientific theories. Since theories are often taken to be attempts to formulate the laws of nature, van Fraassen must come up with an alternative account.

His solution lies in a particular interpretation of the semantic conception of scientific theories. Although theories are often presented as (more or less) axiomatized systems from whose axioms the models can be determined, van Fraassen sees the axioms or formulation of laws as merely accidental artefacts.

---

[10] Except for this: The connection between symmetries and conservation laws is a consequence of Noether's theorem, which shows that under certain conditions every symmetry of a Lagrangian implies a conserved current. But Noether's results are only relevant for systems governed by a Lagrangian, i.e. systems that obey the Euler–Lagrange equations, and this is a matter of being governed by certain sorts of laws.

The content of the theory is not in the laws but in the set of models one ends up with: 'if the theory as such is to be identified with anything at all—if theories are to be reified—then a theory should be identified with its class of models' (van Fraassen 1989, p. 222).

Van Fraassen adds to this view his own peculiar account of the objectives of theorizing. Famously, he believes that science is not a search for theories which are true, i.e. which contain a model which correctly describes the entire actual world, but rather theories that are empirically adequate, i.e. which contain a model which fits all of the observable phenomena in the actual world.

Now science cannot aim *merely* at theories that are empirically adequate, for that is trivially accomplished. The 'anything goes' theory has as models all logically possible distributions of observable phenomena through space-time, and so can accommodate itself to anything we see. Such a 'theory' would not explain anything, and could not be used to predict anything, but it would be empirically adequate.

For the same reason, science cannot seek merely theories that are *ontologically adequate*, i.e. that contain a model that is isomorphic to the whole actual world, observable and unobservable. We seek theories that are small classes of models; at least we want to cut down the models of the 'anything goes' theory. The questions are how this is done and why this is done. What are the principles that govern the acceptance of ever stronger theories?

One might maintain that we seek theories that are *metaphysically adequate*, i.e. theories whose models stand in one-to-one correspondence with the physically possible states of affairs, each model being isomorphic to a state. On the view that we seek theories that correctly state the laws of nature, this is true. But it is unavailable to van Fraassen unless he reifies the physically possible worlds, and is unavailable on general empiricist grounds even then. So we are still left with the questions: how do we slim down our theories and how does their explanatory power increase in virtue of reducing the class of models?

If a theory is just a class of models then the only solution that appears (to me) to be possible is that a theory explains a fact in the actual world only if that fact obtains in all (or an appropriately defined most) of its models. We can state this position without reference to laws and hence without reference to explanation by nomic subsumption. But it is easy to see that if we let the models play the role of Vallentyne's world-histories then we can adopt his definition of law: a law is a proposition that is true in all the models. Then the proposed account of explanation by theories amounts to explanation by

A Modest Proposal 39

subsumption under Vallentyne-laws. Van Fraassen would have to regard this way of putting the matter as innocuous since we have defined Vallentyne-laws from his primitives. We now have a rationale for improving on the 'anything goes' theory: theories with fewer models can explain more.[11] But this account buys explanation too cheaply. For we must take quite seriously the idea that the theory *is just* the class of models. The mode of presentation, the 'derivation' of the models from laws and boundary conditions and choices of results for stochastic processes, is inessential. Consider the theory that results if we begin with our present theories of physics, chemistry, etc. but exclude all the models which do not contain living beings. This is a perfectly good class of models, and hence a theory. In this theory, the existence of life in the universe is a 'law'. Hence the existence of life in the universe is explained as the consequence of a 'law'. If our old theory was empirically adequate or ontologically adequate, so is the new one. Hence the new theory is preferable to the old one: it is as likely to be adequate as the old and more explanatory. And clearly the same trick can be played for any known fact: simply restrict the class of models from the old theory to those in which the fact holds, and thereby produce a theory that explains it. This does not describe actual practice.

Since science would not rest content with the 'anything goes' theory, the drive for, and constraints on, the search for logically stronger theories must be explained. It is hard to see how to explain this drive if one takes the class of models as primitive and the means of presenting or defining the class as incidental. On the view defended in this essay models are derivative, laws primary. The content of a particular model is determined by three factors: the laws, the boundary values, and the results of stochastic processes. Correspondingly, there are three kinds of regularity in a model: those due entirely to the laws, those due in part to the boundary conditions, and those due in part to the results of stochastic processes (the last two classes may overlap). Regularities of the first kind are explicable by nomic subsumption, the other kinds are accidental and have no explanation.[12] There is no explanation in the stochastic 'all heads' world for the regularity of results of coin flips. It is just by chance.

---

[11] I have here ignored much of van Fraassen's account of the pragmatics of explanation in his (1980), but they are not relevant for the point. Once the various context-dependent features of the explanatory context have been determined, one still must explain why logically stronger theories can explain more than logically weaker ones.

[12] This formulation is too strong, but is a good starting point. As we will see below, there are degrees to which a regularity can depend on particular initial conditions or on fortuitous results of random processes. If the degree is sufficiently reduced we consider the regularity explained.

So far we have dealt only in toy cases. But scientific practice supplies striking examples of the search for explanation of regularities by nomic subsumption and the role that search plays in evaluation of theories. The next section will put some meat on these logical bones by examining a contemporary example.

7. *EXEMPLI GRATIA*: THE INFLATIONARY COSMOLOGY[13]

If one looks into the sky in the microwave range one sees, in all directions, an extremely uniform background radiation with an effective temperature of about 3° Kelvin. The temperature of the radiation deviates no more than one part in 10,000 in different regions of the sky. If one calculates the ratio of the energy density of the universe to the critical density, i.e. the density which divides opens universes (which continue expanding forever) from closed universes (which end in a Big Crunch), one finds that it is near 1: current calculations yield a range of values from 0.1 to 2. If one looks for magnetic monopoles, massive particles with an isolated magnetic pole, one does not readily find one; indeed no uncontroversial sighting has yet been reported. If there are any, there are few.

Standard Big Bang cosmology is ontologically (and hence empirically) adequate with respect to these facts. There are models of Big Bangs which have as uniform background radiation as one likes, that have energy densities arbitrarily close to critical density, that have few or no magnetic monopoles. Nonetheless, these phenomena are considered to pose deep, perhaps fatal, objections to standard Big Bang cosmology. Why? To avoid the suspicion that a philosophical interpretation is being imposed on this case study, I quote at length the cosmologists Alan Guth and Paul Steinhardt:

When the standard big-bang model is extended to these earlier times [i.e. between 10^{-43} and 10^{-30} seconds after the singularity], various problems arise. First, it becomes clear that the model requires a number of very stringent assumptions about the initial conditions of the universe. Since the model does not explain why this set of initial conditions came to exist, one is led to believe that the model is incomplete—a theory which included a dynamical explanation for these conditions would certainly be more convincing. In addition, most of the new theories of elementary particles imply that exotic particles called magnetic monopoles (each of which corresponds to an isolated north or south magnetic pole) would be produced in the early universe.

---

[13] This presentation follows that of Guth and Steinhardt 1989, which is highly recommended to those interested in a non-technical presentation of the physics.

In the standard big-bang model, the number of monopoles would be so great that their mass would dramatically alter the evolution of the universe, with results that are clearly inconsistent with observations.

The inflationary universe model was developed to overcome these problems. The dynamics that govern the period of inflation have a very attractive feature: from almost any set of initial conditions the universe evolves to precisely the situation that had to be postulated as the initial state in the standard model. Moreover, the predicted density of magnetic monopoles becomes small enough to be consistent with the fact that they have not been observed. In the context of the recent developments in elementary particle theory, the inflationary model seems to be a simple and natural solution to many of the problems of the standard big-bang picture. (Guth and Steinhardt 1989, p. 34)

Why does the standard theory have such problems? First microwaves. According to the standard scenario the universe expanded from the initial singularity at a relatively constant rate (on the appropriate logarithmic scale). If we retrodict using this model of expansion we find that the regions from which the microwaves originated were never in causal communication with each other (if no cause operates at greater than the speed of light). These regions could therefore never have come into thermal equilibrium. The radiation that comes to us from the north traces back to a different part of the initial boundary conditions than that which comes from the south, and the processes by which the radiation has propagated have never been in causal contact. So the only way for the radiation to have precisely the same temperature in both places is for the initial boundary conditions to be already highly uniform. Of all the distributions of values that the independent variables in the initial state could take, they just happened to take an almost perfectly uniform distribution. Thermal equilibrium could account for uniformity over small patches of the sky, but uniformity over the whole sky must be put in by hand, as a very special choice of initial conditions.

Similarly, the energy density of the universe must be put in by hand. In the standard model this parameter is a free variable that could take any value. Furthermore, in the standard dynamics the critical density is a point of unstable equilibrium.

If W [the ratio of the energy density of the universe to the critical energy density] was ever exactly equal to one, it would remain exactly equal to one forever. But if W differed slightly from one an instant after the big bang, then the deviation would grow rapidly with time. Given this instability, it is surprising that W is measured today as being between 0.1 and 2. (Cosmologists are still not sure whether the universe is open, closed, or flat.) In order for W to be in this rather narrow range today, its value after the big bang had to equal one within one part in 10^{15}. The standard model offers no explanation of why W began so close to one, but merely assumes the fact as an initial condition. (ibid. 37)

Finally, magnetic monopoles. According to present theory, the universe contains a field, the Higgs field, with the following peculiar property. When the field is uniformly zero (the 'false vacuum'), it is not in its lowest energy state. The false vacuum can decay down into a lower energy state (the 'true vacuum'), liberating energy in the process. In the true vacuum, the Higgs field takes on a non-zero value. Indeed, there are *many* true vacuum states, in each of which the Higgs field has a *different* value. The process of decay is called *spontaneous symmetry breaking* since the Higgs field goes from a symmetrical false vacuum state to a particular true vacuum that does not display symmetry under transformations of the field. The process of spontaneous symmetry breaking is random, being governed by quantum laws. We cannot predict when it will occur or, more importantly, into which true vacuum state the field will decay.

After the Big Bang, the Higgs field will be in a false vacuum state. As the universe cools, it becomes energetically favorable for spontaneous symmetry breaking to occur. Once the field decays at a certain point, that value of true vacuum can spread out from the point, but again not faster than light speed. So the same problem as appeared with the radiation appears again. There can be no mechanism to ensure that regions that are not in causal communication end up in the same true Higgs vacuum state. Regions outside the light-cone 'horizon' of one decay event will decay independently, and it is almost unimaginably unlikely that they will arrive at the same true vacuum state.

What happens when two regions with different true vacuums collide? The incompatibility of the vacuum states causes defects to appear where the Higgs field is not well-defined. Linelike defects are called *vortices* or *cosmic strings*; planelike defects are *domain walls*. Pointlike defects are *monopoles*. The more independent decay events there are, the more defects of all kinds there should be. According to the standard cosmology, many decay events should have occurred because the visible universe evolved from many causally unconnected regions. But few if any monopoles exist.

To repeat, it is not that the standard cosmology cannot accommodate the observed facts. We can choose initial conditions that yield smooth and uniform background radiation; we can set W as near as we like to one; we can even assert that the various random decay events all happened to decay to the same value, or that there has only been one decay event, so there are no defects. But these occurrences are monstrously unlikely. Most initial conditions give a universe far from the critical energy density with widely varying background radiation.[14] The quantum mechanical probability of having unnoticeably few monopoles is incalculably small. Despite all that, the standard model is ontologically adequate.

But as Guth and Steinhardt insist, although the standard cosmology can *accommodate* these facts, it cannot *explain* them. The observed regularities would be due to primitive regularities in the initial data or to remarkable fortuitous results of random processes, and this does not constitute an explanation. We want a *dynamical* explanation, a theory that traces these regularities back to the operation of laws. The phenomena should appear independently (or nearly independently) of initial conditions and should not need conspiracies among random processes.

Roughly, here is how the inflationary theory accomplishes this. Recall that we don't have any trouble explaining why a region of the visible universe that traces back to a small enough segment in the early universe has uniform radiation or few monopoles. In a small enough segment, the different places are in causal contact with one another. So the radiation can come into thermal equilibrium and the Higgs vacuum can spread to fill the whole space. The problem is that in the standard picture the visible universe expanded from many causally unconnected regions, not from one connected one.

On the inflationary model, the universe at 10^{-43} seconds was at least some fifty orders of magnitude smaller than on the standard picture, so small that the whole of the visible universe could trace back to one causally connected region. At about 10^{-35} seconds a period of hyper-exponential inflation set in, driven by the decay from the false vacuum to the true vacuum. Between 10^{-35} and 10^{-33} seconds the universe grew by a factor of 10^{50}. After that the Higgs field freezes out in the true vacuum and standard Big Bang cosmology takes over.

As a result, the microwave and monopole problems are solved. The distant ancestors of the present microwave fields were in thermal equilibrium, so they had to come to the same temperature. Only very few events of spontaneous symmetry breaking occurred. And as a bonus, the hyperinflation tends to 'smooth out' the curvature of the universe, driving it towards the critical value of energy density rather than away. From almost any initial conditions

---

[14] This 'most' requires some measure over the space of initial conditions. One can squabble about the exact form of a measure, but for any reasonable, intuitive measure the space of conditions leading to the observed regularities is minuscule.

the dynamical laws deliver a universe nearly flat, with constant radiation, and with few defects.

It is clear in what sense the inflationary model explains the features that the standard model does not. The phenomena do not follow from the laws alone, but in the inflationary model *avoiding* the observed results requires the same fine-tuning of initial conditions, the same fortuitous results of random process, that the standard model needs to *accommodate* the phenomena. The inflationary laws render the results massively insensitive to changes in initial data and give them a high objective probability. It is fair to call this explanation by subsumption under law.

We can understand this judgement of explanatory power if we recognize the role of laws in generating models. The data are explained by the laws because the data obtain in most of the models generated by the laws together with the variety of possible initial conditions and results of random processes. In most of the physically possible universes that develop according to these laws, the radiation is uniform, etc. The effects are explained by the laws because they occur in most of the models generated by the laws.

But we cannot understand these judgements of explanatory power on the semantic view as van Fraassen espouses it. If we care only about the set of models of the theory and not about the laws that generate them, then the only way in which the inflationary theory can be said to explain the data is that the data occur in most of the models. But if that is all that was needed, an explanatory theory was always near to hand. Simply take the models of the standard cosmology and throw out all the models with non-uniform radiation, with energy densities far from the critical value, or with many magnetic monopoles. In this new theory, this new class of models, the troubling phenomena occur in most of the models. But this ridiculous ad hoc procedure would not provide a theory which explains the phenomena. Indeed, if one accepts a semantic view according to which formulations of laws are only inessential means of specifying classes of models, no sense at all could be made of the demand of Guth and Steinhardt for a *dynamical* explanation. The dynamics depend on the laws, so an account of scientific practice that can accord with this episode must allow laws a non-incidental role in theories.

8. PERORATION

In many respects this proposal about laws of nature is modest in just the way Swift's was. I have urged a radical reconstruction of the problematic surrounding laws, shifting them from analysandum to analysans. With laws in hand, connections to possibilities, to counterfactuals, and to explanations come swiftly and easily, with better results than many other philosophical approaches yield. But in other ways, this proposal is truly modest, a sketch to be elaborated. I would like to indicate some directions I think this elaboration should take, and some respects in which modesty will ever remain a virtue.

This chapter has concentrated on *fundamental* laws of temporal evolution, with examples drawn from physics. The mark that we believe a law to be a fundamental law is that exceptions are not tolerated. If the law describes the pattern that nature respects in generating the future from the past then those patterns must appear wherever the law is in force.

The laws of the special sciences do not aspire to this status. Biology, chemistry, psychology, geology, as well as our everyday rules of thumb about how things go, all involve descriptions of how systems typically evolve, but exceptions of certain sorts are both tolerated and expected. Not every embryo follows the usual developmental pattern, not every gene pool evolves according to the laws of population genetics. To take the most blatant example, the laws of population genetics will not govern the evolution of a species subject to nuclear holocaust. An acid mixed with a base will not form a salt if the student knocks the beaker off the table.

Laws that admit of exceptions are sometimes called *ceteris paribus laws*, and are seen as posing philosophical difficulties. The term 'ceteris paribus' is inapplicable here, though, absent any reference class in comparison with which we can determine whether the *cetera* are *paria*.[15] Since these are typically laws of the special sciences, let us call them instead special laws. Some sciences, such as population genetics, contain both Special Laws of Temporal Evolution (SLOTEs) and special adjunct principles.

What is the logical form of a special law? How do special laws connect with possibility, counterfactuals, and explanation?[16]

My proposal is this: the logical form of a special law is just like the logical form of a fundamental law with the addition of the clause 'if nothing interferes'. Population genetics, if true, tells us how gene pools will (or might) evolve if nothing interferes. The connections to possibility, counterfactuals,

---

[15] We may contrast this with the *ceteris paribus* condition that applies at Step 2 of the recipe: there the actual world provides the standard of comparison. Of course, we must still decide which *cetera* are to be kept *paria*.

[16] The following thoughts on special laws have been greatly influenced and stimulated by discussion with Brian McLaughlin.

and explanations are taken over *in toto* to the special sciences. The laws of population genetics beget models that describe how gene pools evolve if nothing interferes. They support counterfactuals about what would or might happen in various circumstances if nothing interferes. They explain features of genetic evolution when nothing interferes. And if the laws of population genetics are true, then actual populations will evolve in accord with them if nothing interferes.

Now all of our suspicions have become focused on the rider 'if nothing interferes'. One line of thought maintains that as it stands the rider is empty; it can only be given content if all of the possible forms of interference are explicitly enumerated. But now the laws of chemistry or biology must become bloated with admonitions about nuclear explosions, meteor collisions, clumsy students, etc. And no matter how bloated it gets, some possible interfering factors will always be left out.[17]

One further suspects that 'if nothing interferes' can be used as a weasel clause to save bad theories. Telepathy is alleged to be a highly fragile phenomenon with everything from sunspot to skeptics producing bad vibes that destroy the effect. Isn't such an escape clause the mark of pseudo-scientific wimps who won't face the tribunal of experience?

And lastly, one might maintain that such a clause robs a theory of its explanatory power. Traditional theories of explanation discerned a parallelism between explanation and prediction. But if the content of the interference clause is not made explicit we can no longer predict anything with confidence. We won't know if we have controlled for all possible forms of interference. The best we can do is to make a conditional prediction: if nothing interferes, then ...

This last objection surely depends on a criterion of explanation that is too strong, for it would rule out most scientific practice. No chemist would predict the result of an experiment if we admonish her to take the likelihood of nuclear war into account. No psychologist would be surprised if a subject failed to act as predicted if in the interim the subject had been struck by lightning; nor would the psychologist feel obliged to take up meteorology. Telepathic phenomena may claim to be fragile, but so is scientific instrumentation. If the experimental reports of one lab are not verified by a second lab following the

---

[17] In some cases, all interference may be eliminated by restricting the law to hold for *isolated systems*, where an isolated system is one surrounded by a surface through which no energy or matter is allowed to pass. Unfortunately, no actual systems are isolated in this way.

most exact instructions we can concoct, we do not automatically reject our theory. There might be a bug in their apparatus, a bug of a sort that is not familiar.

No scientist will predict with certainty that a new instrument will work as it was designed to. But if it does so work, its behavior can be explained by the laws used to design it. The original prediction tacitly contains the rider: such-and-such will happen (if nothing interferes). If such-and-such *does* happen, and is not antecedently likely (e.g. if a sharp image of a binary star system appears), then one infers (fallibly) that nothing did interfere. And if nothing interfered, the whole battery of special laws shed their interference clauses and can be used for explanation as if they were fundamental.

The important case is that in which the instrument fails to behave as predicted. Since the laws imply that it will so behave if nothing interferes, if it doesn't work (and we still believe the laws) we infer that something interfered. The pseudo-scientist stops here, taking credit for the successes and fobbing off the failures. But this attitude is not warranted by the law. The conclusion that something interfered is a contingent claim, and as such we must seek to verify it. A bug in the apparatus ought to be able to be found and corrected. If sunspots caused the problem then shielding should eliminate it. If there was interference then there must be some event or object such that had it been removed the expected result would have ensued. This last counterfactual is, of course, underwritten by the FLOTEs.

We have already seen that some pragmatic considerations beyond the laws themselves enter into the evaluation of counterfactuals. To this extent, SLOTEs may be less objective than FLOTEs, for the SLOTE is true only if counterfactuals about the removal of interfering circumstances are true. Furthermore, there may be perfect agreement about under what circumstances the SLOTE would obtain (i.e. give the predicted result) but still disagreement about which of those circumstances constitute the presence or absence of interfering factors.

In many cases, such as that of the student knocking the beaker off the table, it is easy to locate uncontroversial events that constitute interference. And there are equally clear cases where interference does not exist. Suppose one were to attempt to defend as a psychological law the claim that all people, when thirsty, stand on their heads (if nothing interferes). Or worse, that they float to the ceiling (if nothing interferes). These laws are confronted by an uncounted multitude of prima facie counter-examples. It would be safe to say that there are no factors that would be considered interfering factors such that, had they been removed, the predicted consequent would have ensued.

So special laws are not reduced to being empty by the interference clause, even though in some cases the application of the clause may be moot. If this approach is anywhere near being on the right track, the conditions under which we judge that interference has occurred are worthy of further study.[18]

The deepest sense in which this essay has been modest is the degree to which central notions, such as 'law of temporal evolution', 'transmission of infection', and now 'interference', have been left intuitive and vague. There are clear examples of the application of these concepts, and in exactly the cases where these concepts have clear application our intuitions about possibility, counterfactuals, and explanations are clearest. This suggests we are on the right track. But I have provided no rules for the extension of these notions to new contexts.

To some extent we are in a position to do this. For example, our paradigms of FLOTEs, Schrödinger's equation, and Newton's laws are defined for non-relativistic contexts. The idea of a law governing temporal evolution can be extended to special relativistic, and (I think) even general relativistic, contexts. But no philosopher could have predicted these theories before their formulation and allowed for them. Future theories may challenge the notion of time itself, or of a physical state. If space-time becomes a quantum foam at small length scales, the idea of temporal evolution may break down there. But it is the role of philosophers to study and learn from the radical reconceptions of physical reality, not to dictate to them. This sort of vagueness about notions such as 'law of temporal evolution' is, if not a virtue, at least a necessary evil.

Some possible laws would occasion revisions to the explications offered in this paper. For example, a temporal-action-at-a-distance law, according to which some events influence events at later times without having any effect on the intervening physical states, would entail a reworking of the idea of a Cauchy surface, and a corresponding change in the three-step recipe. More radical suggestions may yield situations where the recipe is inapplicable even with revision, and must be given up.

But the mark of a successful philosophical account of the sort offered here is not the universal applicability of its primitive concepts under every circumstance. The mark is rather that when the primitive concepts fail or become fuzzy, so should intuitions about all the other notions that have been built on them. One last example may illustrate this point.

---

[18] Such a study would doubtless begin by delineating *degrees* of interference (and hence degrees of isolation). Probably any actual biological process is affected to some degree by interfering events (e.g. the passage of cosmic rays). Usually such events only slightly perturb the system and so can be ignored if one is interested in gross phenomena.

One analysis of counterfactuals in stochastic contexts relied on the idea of infection. We supposed that the physical magnitudes that are changed at Step 2 can be identified and the subsequent physical magnitudes that depend on them distinguished. This picture is clearly applicable when the physical state consists in locally defined physical magnitudes and the laws are laws of local interaction. In that case, the infection spreads out from an initial locus and can be traced.

In quantum theory both of these conditions may fail. The famous 'entangled states' of a correlated pair of particles cannot be understood as the sum of local states of each particle. The laws describe the evolution of a vector in Hilbert space, not the evolution of spatially local magnitudes. And on the Feynman path integral picture, every particle travels along every possible trajectory, so infection would immediately spread everywhere.

And in exactly the cases of correlated quantum pairs, intuitions about counterfactuals break down. The disputes over the so-called 'Principle of Counterfactual Definiteness', used by Stapp and Eberhard in their proof of Bell's theorem, illustrate this (see Redhead 1987, pp. 90 ff.). The weirdness of quantum theory arises in part from its introduction of concepts that defy easy analysis in terms of the notions that underpin our judgements of counterfactuals, locality, and causality.

The notions of law, possibility, counterfactual, and explanation are deeply interconnected. The directions of the connections ought to be open to all hypotheses. If we take laws as primitive, relatively clean analyses can be given of the rest, analyses that fit intuitions, predict degrees and sorts of context dependence, and describe actual scientific practice, at least in some notable instances. Shortcomings of other theories are avoided and their failures explained. Of course, a polemical essay such as this contains only the happiest examples and competing theories are not given the benefit of every doubt. But I hope that at least a first foothold has been carved for the idea that laws are primitive, primitive components of our conceptual system and even primitive elements of physical reality.

2

Why Be Humean?

Everything only connected by 'and' and 'and'.
(Elizabeth Bishop, *Over 2,000 Illustrations and a Complete Concordance*)

The title of this chapter is not a rhetorical question. Nor is it a question that the essay aspires to answer. It is, rather, a sincere plea for enlightenment. There is, in some contemporary metaphysics, an explicit preference, or desire, or in some cases demand, for 'Humean' theories. Humean, or 'empiricist', theories of law and of chance are sought; theories that posit irreducible nomic or modal or dispositional or causal facts are dismissed as unHumean. David Lewis has characterized a central motivation for some of his theories as a desire to 'uphold not so much the truth of Humean supervenience but the *tenability* of it' (Lewis 1986a, p. xi), a somewhat modest but still mysterious ambition. Why, to put it bluntly, should one *want* to be Humean? What is the appeal of 'Humean Supervenience' such that metaphysical accounts should aspire to it? Although Lewis and others issue calls to rally to Hume's banner, no strategic justification for this campaign is offered. I suspect that the reason for this reticence is that the motivations will not withstand close scrutiny in the light of day. The aim of this essay is to unshutter the windows.

Any examination of Humean Supervenience must begin with a statement of the doctrine. Lewis provides the *locus classicus*:

Humean supervenience is named in honor of the greater [sic] denier of necessary connections. It is the doctrine that all there is to the world is a vast mosaic of local matters of fact, just one little thing and then another. (But it is no part of the thesis that these local matters of fact are mental.) We have geometry: a system of external relations of spatio-temporal distance between points. Maybe points of spacetime itself, maybe point-sized bits of matter or aether fields, maybe both. And at those points we have local qualities: perfectly natural intrinsic properties which need nothing bigger than a point at which to be instantiated. For short: we have an arrangement of qualities. And that is all. All else supervenes on that. (ibid., p x)

and again:

The question turns on an underlying metaphysical issue. A broadly Humean doctrine (something I would very much like to believe if at all possible) holds that all the facts there are about the world are particular facts, or combinations thereof. This need not be taken as a doctrine of analyzability, since some combinations of particular facts cannot be captured in any finite way. It might better be taken as a doctrine of supervenience: if two worlds match perfectly in all matters of particular fact, they match in all other ways too – in modal properties, laws, causal connections, chances .... (ibid. 111)

Although he does not remark it, Lewis's Humeanism comprises two logically independent doctrines. The first, which we may call *Separability*, claims that all fundamental properties are local properties and that spatio-temporal relations are the only fundamental external physical relations. To be precise:

Doctrine 1 (Separability): The complete physical state of the world is determined by (supervenes on) the intrinsic physical state of each spacetime point (or each pointlike object) and the spatio-temporal relations between those points.

Separability posits, in essence, that we can chop up space-time into arbitrarily small bits, each of which has its own physical state, much as we can chop up a newspaper photograph into individual pixels, each of which has a particular hue and intensity. As the whole picture is determined by nothing more than the values of the individual pixels plus their spatial disposition relative to one another, so the world as a whole is supposed to be decomposable into small bits laid out in space and time.

The doctrine of Separability concerns only how the total physical state of the universe depends on the physical state of localized bits of the universe. The second component of Lewis's Humeanism takes care of everything else:

Doctrine 2 (Physical Statism): All facts about a world, including modal and nomological facts, are determined by its total physical state.

I have employed the unlovely neologism 'Physical Statism' to distinguish Doctrine 2 from Physicalism. Physicalism holds that two worlds that agree in all physical respects (i.e. with respect to all items that would be mentioned in a perfected physics) agree in all respects. Physicalism is a much less contentious thesis than Doctrine 2. Doctrine 2 essentially adds to Physicalism the further requirement that all physical facts about the world are determined by its total physical state, by the disposition of physical properties. If one holds,[1] for example, that the laws of nature do not supervene on the total physical state of the world (at least so far as that state can be specified independently of the laws), then one can be a Physicalist while denying Physical Statism. One can hold that worlds which agree on both their physical state and their physical laws agree on all else, while denying that the laws are determined by the state. Lewis's Humeanism importantly maintains the stronger claim.

In order to clearly distinguish Doctrine 2 from Physicalism, we must remark a condition on acceptable analyses accepted by the Physical Statist but not by the Physicalist:

> Non-circularity condition: The intrinsic physical state of the world can be specified without mentioning the laws (or chances, or possibilities) that obtain in the world.

When allied with the doctrine of Separability, the non-circularity condition implies that the physical state of every space-time point is metaphysically independent of the laws that govern the world. This in turn implies that the fundamental physical quantities, such as electric charge, mass, etc., are metaphysically independent of the laws of electromagnetism, gravitation, and so on. This is a controversial thesis, but one that Lewis accepts. It will not come in for further notice here.

The conjunction of Separability with Physical Statism is not peculiar to Lewis. Consider John Earman's account of what it is to be an empiricist about laws of nature:

The filling I prefer for the blank in (F1) produces the following constraint:

(E1) For any W_1, W_2, if W_1 and W_2 agree on all occurrent facts, then W_1 and W_2 agree on laws.

I will refer to (E1) as the empiricist loyalty test on laws, for I believe it captures the central empiricist intuition that laws are parasitic on occurrent facts. Ask me what an occurrent fact is and I will pass your query on to empiricists. But in lieu of a reply, I will volunteer that the paradigm form of an occurrent fact is: the fact expressed by the sentence $P(o, t)$, where 'P' is again a suitable kosher predicate, 'o' denotes a physical object or spatial location, and 't' denotes a time. There may also be general occurrent facts (I think there are), but these are also parasitic on the singular

---

[1] Cf. 'A Modest Proposal Concerning Laws, Counterfactuals, and Explanations', Chapter 1, this volume; Carroll 1994.

occurrent facts. Conservative empiricists may want to restrict the antecedent of (E1) so as to range only over observable facts while more liberal empiricists may be happy with unobservable facts such as the fact that quark q is charming and flavorful at t. In this way we arrive at many different versions of the loyalty test, one for each persuasion of empiricist. (Earman 1984, p. 195)

Note that Earman's 'paradigm form' of an occurrent fact is a fact about a particular (presumably small) space-time region, thus endorsing Separability. Note also that among the particular occurrent facts about a small space-time region there had best not be facts about the laws which govern that region, else an empiricist analysis of laws could consist in stating, independently of all other facts, which laws obtain at every region of space-time.

The interest in dissecting Humean Supervenience into Separability and Physical Statism arises, in the first place, from the remarkable fact that contemporary physics strongly suggests that the world is not Separable. If quantum theory is even remotely on the right track, then the best physical theories will continue, as they do now, to posit fundamental non-Separable physical states of affairs. This discovery casts the question of *motivating* a desire to defend Doctrine 1 into a peculiar light, for one knows beforehand that the motivations, whatever they may be, turn out to lead away from the truth. So before asking why one might want to be Humean, we shall review the evidence that the world is not Humean. Only then will we seek the motivations for defending Separability, and then lastly turn to the possible motivations for Physical Statism.

1. NON-SEPARABILITY IN QUANTUM THEORY

The notion that the physical state of the world is separable is not a philosopher's fancy. In a now famous letter to Max Born, Albert Einstein stated the doctrine succinctly and lucidly:

If one asks what, irrespective of quantum mechanics, is characteristic of the world of ideas of physics, one is first of all struck by the following: the concepts of physics relate to a real outside world, that is, ideas are established relating to things such as bodies, fields, etc., which claim 'real existence' that is independent of the perceiving subject-ideas which, on the other hand, have been brought into as secure a relationship as possible with the sense-data. It is further characteristic of these physical objects that they are thought of as arranged in a space-time continuum. An essential aspect of this arrangement of things in physics is that they lay claim, at a certain time, to an existence independent of one another, provided these objects 'are situated in different parts of space'. Unless one makes this kind of assumption about the independence of the existence (the 'being-thus') of objects which are far apart from one another in space—which stems in the first place from everyday thinking—physical thinking in the familiar sense would not be possible. It is also hard to see any way of formulating and testing the laws of physics unless one makes a clear distinction of this kind. This principle has been carried to extremes in the field theory by localizing the elementary objects on which it is based and which exist independently of each other, as well as the elementary laws which have been postulated for it, in the infinitely small (four-dimensional) elements of space.

The following idea characterizes the relative independence of objects far apart in space (A and B): external influence on A has no direct influence on B; this is known as the 'principle of contiguity', which is used consistently in the field theory. If this axiom were to be completely abolished, the idea of the existence of (quasi-) enclosed systems, and thereby the postulation of laws which can be checked empirically in the accepted sense, would become impossible. (Born 1971, pp. 170–1)

It is no accident that Einstein discusses this principle in connection with quantum mechanics, for Einstein saw, perhaps earlier than anyone else, that the formalism of the quantum theory seems to reject Separability. Let's review quickly why that is so.

The quantum theory, no matter how interpreted, employs as a fundamental device the so-called *quantum state* or *wavefunction* of a system. These quantum states obey a *principle of superposition* to the effect that if **A** represents one quantum state of a system and **B** another, then $\alpha\mathbf{A} + \beta\mathbf{B}$ represents a third possible state of the system, where α and β are complex numbers such that $|\alpha|^2 + |\beta|^2 = 1$. For our example, we will use the *spin states* of electrons; the particular mathematical details of how these states are represented will be of no concern.

The spin of an electron can be measured in any direction in space, and when measured one is certain to get one of two possible results which correspond, using the usual measurement apparatus, to the particle being deflected either in one direction or in the opposite direction by a magnetic field. So if we want to measure the spin of an electron in the z-direction, we shoot the electron through a particularly oriented device and the electron will either be deflected up or down. In the first case we say the electron has z-spin up, in the second z-spin down. It turns out that there is a unique spin state for an electron in which it is *guaranteed* to be deflected upwards in an z-spin measurement, and such an electron is said to have z-spin up; similarly for z-spin down. Using an obvious notation, we represent the z-spin up state as

$|z\uparrow\rangle$ and the z-spin down state as $|z\downarrow\rangle$. Similarly there are up and down spin states for spin measured in any other direction. Quite significantly, any spin state in any direction can be written down as a superposition of the up and down spin states for any other direction. So, for example, we can write the z-spin states as superpositions of the up and down spin states in the x-direction (i.e. a direction orthogonal to the z-direction) as follows:

$$|z\uparrow\rangle = \frac{1}{\sqrt{2}}|x\uparrow\rangle + \frac{1}{\sqrt{2}}|x\downarrow\rangle$$
$$|z\downarrow\rangle = \frac{1}{\sqrt{2}}|x\uparrow\rangle - \frac{1}{\sqrt{2}}|x\downarrow\rangle.$$

z-spin up is a superposition of x-spin up and x-spin down, as is z-spin down.

We can now introduce the only rule of quantum mechanical calculation we will need. Suppose an electron is in some arbitrary spin state **A**, and we decide to measure its x-spin. The quantum formalism tells us to calculate the probabilities of up and down results in the following way. First, write down the state **A** in terms of the states $|x\uparrow\rangle$ and $|x\downarrow\rangle$, that is, write **A** as a superposition of x-spin states $\alpha|x\uparrow\rangle + \beta|x\downarrow\rangle$. Then the probability of getting an up result when measuring x-spin is just $|\alpha|^2$ and the probability of a down result is $|\beta|^2$. To take a concrete example suppose one measures the x-spin of an electron in the z-spin up state $|z\uparrow\rangle$. Since $|z\uparrow\rangle = 1/\sqrt{2}|x\uparrow\rangle + 1/\sqrt{2}|x\downarrow\rangle$, the chance of getting x-spin up is 1/2, as is the chance for x-spin down. This simple rule is all of the quantum theory we will require.

(Incidentally, these rules illustrate why the so-called Heisenberg Uncertainty Principle is built into the fundamental structure of the quantum theory rather than being imposed as an *additional* constraint on the formalism. There is, for example, no spin state in which one could predict with certainty the result of both a z-spin and an x-spin measurement. For the only states which guarantee a particular z-spin result are $|z\uparrow\rangle$ and $|z\downarrow\rangle$, and each of these is, as a purely mathematical matter of fact, a superposition of x-spin states.)

None of these facts about quantum spin as yet implies anything about Separability, for we have only been discussing the spin state of a single particle. But once we apply these same principles to systems containing more than one particle, we get some startling results.

Let us now consider a system that consists in two electrons. What sorts of spin states are available to it?

There are, to begin, the sorts of uninteresting states that the principle of Separability would lead one to expect. It is possible, for example, that particle 1 and particle 2 should both be in the state $|z\uparrow\rangle$, or particle 1 can be in $|z\uparrow\rangle$ while particle two is in $|z\downarrow\rangle$, or particle 1 can be in $|z\uparrow\rangle$ while particle 2 is in $|x\uparrow\rangle$. Using an obvious notation, these states for the two-particle system are written $|z\uparrow\rangle_1|z\uparrow\rangle_2$, $|z\uparrow\rangle_1|z\downarrow\rangle_2$, and $|z\uparrow\rangle_1|x\uparrow\rangle_2$ respectively. In general, for any spin state that the first particle alone can be in and any state that the second alone can be in, there is a state of the joint system which assigns exactly the first state to the first particle and the second state to the second. These states are called *product states* of the composite system and are, metaphysically at least, as boring as could be. A product state assigns a perfectly determinate spin state to each of the two particles, and the state of the composite is nothing but the logical sum of the states of the components.

If all quantum states of composite systems were mere product states, then quantum theory would pose no threat to Separability. But from the principle of superposition it follows that there are more possibilities available for the composite system than product states. Since any pair of states of a system can be superposed to yield a new state, any pair of product states can be superposed. For our purposes, we will consider only two such states: the Singlet State and the m = 0 Triplet State:

$$\text{Singlet State}: \quad \frac{1}{\sqrt{2}}|z\uparrow\rangle_1|z\downarrow\rangle_2 - \frac{1}{\sqrt{2}}|z\downarrow\rangle_1|z\uparrow\rangle_2$$

$$\text{m = 0 Triplet State}: \quad \frac{1}{\sqrt{2}}|z\uparrow\rangle_1|z\downarrow\rangle_2 + \frac{1}{\sqrt{2}}|z\downarrow\rangle_1|z\uparrow\rangle_2.$$

Let's play with these states a bit to see what sorts of statistical properties they display.

Suppose we decide to measure the z-spins of both particles. If the system were in the state $|z\uparrow\rangle_1|z\downarrow\rangle_2$ we would get an up result for particle 1 and down for particle 2. If it were in $|z\downarrow\rangle_1|z\uparrow\rangle_2$ we would get down for 1 and up for 2. So in the Singlet State there is a 50 per cent chance of up-down and a 50 per cent chance of down-up, and similarly for the m = 0 Triplet State.

Suppose we decide simply to measure the z-spin of particle 1, and either measure spin in some other direction or nothing at all on particle 2. There is a fundamental quantum mechanical principle that applies here: the statistical predictions for the results of measurement on one part of a composite system are unchanged by conditionalizing on the fact that any sort of measurement was made on another part. This principle is sometimes called parameter independence in the literature, and it holds in all interpretations of the quantum theory when applied to the statistical predictions one can derive from the wavefunction alone.[2] So we know that the statistics displayed by particle 1 under a z-spin measurement are unchanged whether z-spin or anything else is measured on particle 2. And we know that if z-spin is measured on particle 2, particle 1 will come out z-spin up 50 per cent of the time. Therefore the prediction for particle 1 (or particle 2) when z-spin is measured is simply 50 per cent chance of up and 50 per cent down. In fact, the statistical predictions for *any* direction of spin for either particle are 50 per cent up and 50 per cent down: in any direction one wishes to choose, each particle in the Singlet State or the $m = 0$ Triplet State has an even chance of coming out up or down.

This last fact is of considerable interest, for *no pure quantum state for a single particle displays these statistics*. For example, in the state $|z\uparrow>_1$, particle 1 is certain to yield an up result if z-spin is measured. And in every pure single particle quantum spin state, there is some direction such that the result of a spin measurement in that direction is certain. It follows that the Singlet and $m = 0$ Triplet States *cannot be written as simple product states but only as superpositions of product states*. Such states are called *entangled states*, and they engender the most shocking and radical metaphysical innovations to be found in the quantum theory.

The central challenge that quantum theory poses for Separability can now be stated. Suppose there are two electrons, well separated in space (perhaps at opposite ends of a laboratory), that are in the Singlet State. If the principle of Separability held, then each electron, occupying a region disjoint from the other, would have its own intrinsic spin state, and the spin state of the composite system would be determined by the states of the particles taken individually, together with the spatio-temporal relations between them. But, as we have seen, no pure state for a single particle yields the same predictions as the Singlet State, and if one were to ascribe a pure state to each of the electrons, their joint state would be a product state rather than an entangled state. The joint state of the pair simply cannot be analyzed into pure states for each of the components.

---

[2] The reason for the careful qualifications in this sentence is that not all interpretations of the quantum theory display parameter independence at the level of all fundamental ontological posits. Bohm's theory, for example, violates it for individual systems, but still obeys it at the level of statistical predictions. Any theory that disobeyed parameter independence at the level of the statistical predictions from the wavefunction would also violate the so-called quantum no-signaling theorems, and would permit superluminal signals.

The attentive reader will have noticed that the qualifier 'pure' has snuck into the preceding discussion. No *pure* state for an individual particle yields 50–50 chances for spin measurements in all directions; if each particle is in a *pure* state then the pair is in a product state, and so on. And indeed, the qualifier 'pure' is needed because there is another set of states, the so-called *impure states* (or *mixtures* or *statistical operators* or *density matrices*), for which not all of these assertions hold. Roughly speaking, one can think of an impure state as the sort of state one would use to make predictions if one were unsure which pure state a system is in. If, for example, one knew that a single electron were either in $|z\uparrow\rangle$ or in $|z\downarrow\rangle$ but were unsure which, and assigned a 50 per cent chance to each possibility, then one could calculate in the usual way expectations for various measurements. And one would find that in such a state there is a 50 per cent chance for an up or a down outcome for spin measurements in any direction.[3] There is a determinate mathematical procedure for deriving the mixed state of each component of a composite system from the state of the whole, when that state is entangled. The resulting mixed state will make all the right statistical predictions about the component, for all possible measurements performed solely on it.

So why not just say that when a pair of electrons is in the Singlet State each electron is in the appropriate mixed state, and thereby recover Separability? The problem for this approach arises when we consider the $m = 0$ Triplet state. If one makes a z-spin measurement on an electron in the $m = 0$ Triplet State, there is a 50 per cent chance of up and 50 per cent of down, and similarly for measurements in any direction. *The mixed states assigned to component electrons in the Singlet State are identical to the mixed states assigned to component electrons in the $m = 0$ Triplet State.* So if Separability holds, then since each component of a pair of particles in the Singlet State is in exactly the same spin state as each component of a pair in the $m = 0$ Triplet State, and since the spatio-temporal relations between the members of the pair could be identical, the Singlet State would have to be identical to the $m = 0$ Triplet State. Separability holds that the global physical state of a system supervenes

---

[3] The 'roughly' in the foregoing characterization of mixed states is required because the statistical operators or density matrices are less discriminating than states of epistemic uncertainty. If I know that a particle is definitely either in $|z\uparrow\rangle$ or in $|z\downarrow\rangle$, but am completely unsure which, then I am in a different epistemic state from knowing that the particle is either in $|x\uparrow\rangle$ or in $|x\downarrow\rangle$ and being completely unsure which. But both of these epistemic states would yield exactly the same statistical predictions for all observables, and both correspond to the same statistical operator.

Why Be Humean?

on the local states of the parts plus their spatio-temporal relations, and in this case the states of the parts and the spatio-temporal relations (which play no role in the spin state in any case) are identical.

At this point the reader might well wonder why the Singlet State and the m = 0 Triplet State aren't identical. After all, it follows from what has been said that no measurements on a component electron of a pair in the Singlet State can distinguish it from a component of a pair in the Triplet. Taken individually, parts of Singlets act just like parts of m = 0 Triplets. Furthermore, the only difference between the Singlet and the m = 0 Triplet is a minus sign, which seemingly becomes irrelevant when calculating probabilities, since the coefficients in the superposition are squared. But matters are not so simple. Although no *local* measurement on a *single* electron can distinguish the Singlet from the m = 0 Triplet, a *global* measurement on the whole composite system can.

We have already noted that if z-spin measurements are made on both electrons, then the predictions for the Singlet and m = 0 Triplet States are the same: 50 per cent up-down and 50 per cent down-up. It is here that the minus sign disappears in the squaring. But what if we measure the *x*-spins of both particles?

To get the answer, we must express the Singlet and m = 0 Triplet in terms of x-spin. This is possible since the z-spin states can themselves be written as superpositions of x-spin, as already noted. Starting with our original definition of the Singlet and m = 0 Triplet, and substituting

$$|z\uparrow> = \frac{1}{\sqrt{2}}|x\uparrow> + \frac{1}{\sqrt{2}}|x\downarrow> \text{ and}$$

$$|z\downarrow> = \frac{1}{\sqrt{2}}|x\uparrow> - \frac{1}{\sqrt{2}}|x\downarrow>$$

we can derive:

$$\text{Singlet} = \frac{1}{\sqrt{2}} \left(\frac{1}{\sqrt{2}}|x\uparrow>_1 + \frac{1}{\sqrt{2}}|x\downarrow>_1 \right) \left(\frac{1}{\sqrt{2}}|x\uparrow>_2 - \frac{1}{\sqrt{2}}|x\downarrow>_2 \right)$$
$$- \frac{1}{\sqrt{2}} \left(\frac{1}{\sqrt{2}}|x\uparrow>_1 - \frac{1}{\sqrt{2}}|x\downarrow>_1 \right) \left(\frac{1}{\sqrt{2}}|x\uparrow>_2 + \frac{1}{\sqrt{2}}|x\downarrow>_2 \right)$$
$$= \frac{1}{2\sqrt{2}} (|x\uparrow>_1|x\uparrow>_2 - |x\uparrow>_1|x\downarrow>_2 + |x\downarrow>_1|x\uparrow>_2$$

$$-|x\downarrow>_1|x\downarrow>_2) - \frac{1}{2\sqrt{2}}(|x\uparrow>_1|x\uparrow>_2 + |x\uparrow>_1|x\downarrow>_2$$
$$-|x\downarrow>_1|x\uparrow>_2 - |x\downarrow>_1|x\downarrow>_2)$$
$$= \frac{1}{2\sqrt{2}}(2|x\downarrow>_1|x\uparrow>_2 - 2|x\uparrow>_1|x\downarrow>_2)$$
$$= \frac{1}{\sqrt{2}}(|x\downarrow>_1|x\uparrow>_2 - |x\uparrow>_1|x\downarrow>_2)$$

$$m = 0 \text{ Triplet} = \frac{1}{\sqrt{2}}\left(\frac{1}{\sqrt{2}}|x\uparrow>_1 + \frac{1}{\sqrt{2}}|x\downarrow>_1\right)\left(\frac{1}{\sqrt{2}}|x\uparrow>_2\right.$$
$$\left.-\frac{1}{\sqrt{2}}|x\downarrow>_2\right) + \frac{1}{\sqrt{2}}\left(\frac{1}{\sqrt{2}}|x\uparrow>_1 - \frac{1}{\sqrt{2}}|x\downarrow>_1\right)$$
$$\left(\frac{1}{\sqrt{2}}|x\uparrow>_2 + \frac{1}{\sqrt{2}}|x\downarrow>_2\right)$$
$$= \frac{1}{2\sqrt{2}}(|x\uparrow>_1|x\uparrow>_2 - |x\uparrow>_1|x\downarrow>_2 + |x\downarrow>_1|x\uparrow>_2$$
$$-|x\downarrow>_1|x\downarrow>_2) + \frac{1}{2\sqrt{2}}(|x\uparrow>_1|x\uparrow>_2 + |x\uparrow>_1|x\downarrow>_2$$
$$-|x\downarrow>_1|x\uparrow>_2 - |x\downarrow>_1|x\downarrow>_2)$$
$$= \frac{1}{2\sqrt{2}}(2|x\uparrow>_1|x\uparrow>_2 - 2|x\downarrow>_1|x\downarrow>_2)$$
$$= \frac{1}{\sqrt{2}}(|x\uparrow>_1|x\uparrow>_2 - |x\downarrow>_1|x\downarrow>_2)$$

The simple minus sign distinguishing the two states when written in terms of z-spin now looms large: it implies different cancellations when converting to x-spin. And that in turn implies quite different statistics for pairs of x-spin measurements. If the particles are in the Singlet State and x-spin is measured on each, then they are certain to give opposite results, with a 50 per cent chance of up-down and 50 per cent chance of down-up. But the same measurement on the $m = 0$ Triplet State is certain to yield the *same* result for both electrons: half of the time both will be up and half of the time both down. A single global measurement of x-spin is guaranteed to distinguish a pair of particles in the Singlet State from a pair in the $m = 0$ Triplet, so the two states cannot be identified.

The difficulty facing Separability is now inescapable. Consider two pairs of electrons, one in the Singlet and the other in the m = 0 Triplet State, such that the spatio-temporal relations within each pair are identical (e.g. in each pair the electrons are 5 meters apart). Can one analyze the global physical state of each pair into local physical states of each part taken individually plus the spatio-temporal relations? Evidently not. For what is the local physical state of each electron?

If only pure states are allowed as possible physical states, then none of the electrons has any local state, i.e. a state that can be specified without reference to the other member of the pair. One could say that the state of one electron is, for example, being part of a pair of electrons which are in the Singlet State, but that is evidently not a purely local matter. And if none of the electrons has a local intrinsic spin state, the global state of each pair cannot supervene on the local states of the parts plus space-time relations.

If mixed states are allowed as possible physical states of systems, then the problem is still not solved. Each electron can now be assigned its own intrinsic local state, but all four electrons are assigned exactly the same state. So if the global spin state of the system supervenes on the local intrinsic states of the parts plus space-time relations, the two pairs must be in identical spin states, which they are not. Either way you cut it, Separability fails. The upshot is that no physical theory that takes the wavefunction seriously can be a Separable theory. If we have reason to believe that the quantum theory, or any extension of it, is part of a true description of the world, then we have reason to believe the world is not Separable.[4]

2. LEWIS'S REACTION AND THE MOTIVATION FOR SEPARABILITY

Lewis is aware that the quantum theory poses a threat to Separability, and says he is prepared to take the consequences:

Is this [namely Humean Supervenience] materialism?—no and yes. I take it that materialism is metaphysics built to endorse the truth and descriptive completeness of

---

[4] There is one final move that can be made to save Separability, namely demand that one's account of space itself be altered so that the physical state of the world be Separable in the space. Barry Loewer suggests this in 'Humean Supervenience' (1996), recommending that the 'fundamental space' of quantum mechanics be taken to be configuration space, rather than space-time. This constitutes the ultimate elevation of Separability as a regulative principle, rather than an empirical theory, and urges even more strongly the question of motivation.

physics more or less as we know it; and it just might be that Humean supervenience is true, but our best physics is dead wrong in its inventories of the qualities. Maybe, but I doubt it. Most likely, if Humean supervenience is true at all, it is true in more or less the way that present physics would suggest ...

Really, what I uphold is not so much the truth of Humean supervenience but the *tenability* of it. If physics itself were to teach me that it is false, I wouldn't grieve.

That might happen: maybe the lesson of Bell's theorem is exactly that there are *physical* entities which are unlocalized, and which therefore might make a difference between worlds ... that match perfectly in their arrangements of local qualities. Maybe so. I'm ready to believe it. But I am not ready to take lessons in ontology from quantum physics as it now is. First I must see how it looks when it is purified of instrumentalist frivolity, and dares to say something not just about pointer readings but about the constitution of the world; and when it is purified of doublethinking deviant logic; and—most of all—when it is purified of supernatural tales about the power of observant minds to make things jump. If, after all that, it still teaches nonlocality, I shall submit willingly to the best of authority. (Lewis 1986a, p. xi)

If we take Lewis at his word, then we should abandon Separability (and hence his version of Humean Supervenience) forthwith. For one *can* see how quantum physics looks when purified of instrumentalism, and quantum logic, and consciousness-induced wave collapse. This has been done in several quite different ways: in David Bohm's so-called ontological interpretation (see e.g. Bohm and Hiley 1993), in the (mind-independent) spontaneous collapse theories of Ghirardi, Rimini, and Weber (1986) and of Philip Perle (1990), even in the Many Minds theory of David Albert and Barry Loewer (see Albert 1992). These theories all have fundamentally different ontologies and dynamics, but all agree that the physical state of the world is not Separable, for they all take the wavefunction seriously as a representation of the physical state. This is not to say that Non-Separability is absolutely forced on us by empirical considerations: it would not be impossible to construct a Separable physics with the same empirical import as the present quantum theory.[5] But no one is trying to do it, and there seems to be no reason to start: the quantum theory (in a coherent formulation) is elegant, simple, and

---

[5] In this regard, one must carefully distinguish Separability from relativistic Locality, i.e. from the claim that the physical state at any point of space-time is determined or influenced only by events in its past light-cone, or, colloquially, that no influence travels faster than light. Bell's theorem shows that certain empirical predictions of the quantum theory, namely violations of Bell's inequality for events at spacelike separation, cannot be recovered by any local theory. They could, however, be recovered by a Separable theory which contains superluminal or backward causal connections. See my 1994 for all the grisly details.

empirically impeccable. Lewis would not elevate his preference for Separable theories into some a priori constraint which could dictate to physics, as the quote shows. Given the definition of materialism cited above, contemporary materialism (i.e. metaphysics built to endorse the approximate truth and descriptive completeness of contemporary physics) must deny Separability.

This leaves us with two questions. First, what drew Lewis to Separability in the first place? Since the doctrine appears to be false, we ought to consider carefully the grounds upon which it was thought to be established, or at least rendered plausible. Second, and more importantly, what of Physical Statism? This second component of Humean Supervenience remains as yet untouched by any criticism, and one could continue to insist upon it even while abandoning Separability. Perhaps the physical state of the universe does not supervene on the local intrinsic states of its point-like parts together with spatio-temporal relations, but yet the 'modal properties, laws, causal connections, chances' (ibid. 111) all are determined by the non-Separable total physical state of the universe. Perhaps. But our suspicions have been rightly aroused. The considerations in favor of Humean Supervenience already led us astray with respect to Separability, so why think they are likely to be any more reliable with respect to Physical Statism? Before we can even begin to take up this question, we must answer the first: what considerations seemed to support Separability in the first place?

Fortunately, the answer to this question is clear, simple, and intelligible. It has, indeed, already been stated. Lewis wants a metaphysics built to endorse the ontology of physics. And, as the quotation from Einstein above forcefully illustrates, *classical* physics *is* Separable. Classical mechanics and field theory do postulate that the physical state of the whole universe is determined entirely by the spatio-temporal dispositions of bodies, their intrinsic physical properties (such as charge and mass), and the values of fields at all points in space through time. Taking one's ontology from classical physics does entail Separability. But the advent of the quantum theory, as we have seen, has superseded that argument; it is irreparably damaged, and Lewis has nothing more to say.

Perhaps, though, Einstein does. At the end of his discussion, Einstein suggests that Separability is a kind of a priori constraint on any comprehensible and empirically verifiable physics (NB: not an a priori constraint on how the world must be, but how it must be if we are to know it through empirical procedures, a truly Kantian theme). Einstein writes that if Separability 'were to be completely abolished, the idea of the existence of (quasi-) enclosed systems, and thereby the postulation of laws which can be checked empirically in the accepted sense, would become impossible' (Born 1971). It is hard to respond directly to this claim, since no further explanation or justification is offered, but if 'completely abolished' means just 'denied', then the quantum theory itself stands as a refutation of the claim. Quantum theory has both been formulated and rigorously tested despite the centrality of non-Separable elements in its ontology. Whatever Einstein had in mind, he had to be wrong.

(If by 'completely denied' Einstein merely means that any empirically testable theory must postulate *some* local intrinsic physical states, but not that the total physical state of all systems is Separable, then he would be anticipating John Bell's call that one carefully consider what the 'local beables' of a theory are, i.e. the objectively existing quantities which '(unlike the total energy, for example) can be assigned to some bounded space-time region' (Bell 1987, p. 53). One could try to make an argument that a physical theory with *no* local beables cannot be brought into correspondence with our experience of the world, but even this weaker claim may face serious obstacles.)

So no credible motivation for Separability exists in the face of the existence and empirical testability of quantum physics. What of Physical Statism?

3. THE MOTIVATION AND STATUS OF PHYSICAL STATISM

Why believe that the 'modal properties, laws, causal connections, chances' all supervene on the total physical state of the universe, that there could not be two possible worlds which agree on their total physical state but disagree on some of these? If the motivation for Separability is to be found in the ontology of classical physics, and if Lewis's materialism is just metaphysics tailored to endorse the (approximate) truth and descriptive completeness of physics as we know it, then one would first seek the motivation for Physical Statism also in the practices of the physicists (and other natural scientists).

What has physics to say about modal properties, laws, causal connections, and chances? The topic is a large one, so I will treat much of it summarily. Physicists do make assertions about what is physically possible and impossible. Cosmologists, for example, regard both open and closed universes as physically possible, and study the features of each. How they manage this is relatively clear: they have the field equations of General Relativity, and regard the physically possible universes as models of these laws (together, perhaps, with some other conditions, e.g. the absence of closed timelike curves, but this is controversial). So once we have an account of physical law, the account of physical possibility is near to hand. Similarly, physicists are happy to evaluate counterfactuals, so long as they are precisely enough stated. This means that the counterfactual condition must be specified completely enough to delimit a class of models in which the condition is satisfied, and the class is coherent enough that in all (or in most, by some natural measure) the consequent of the counterfactual has the same truth value. So, for example, one asks: what would the Earth be like if there were no moon? *What If the Moon Didn't Exist* by Neil Comins (1995) is devoted to just this question. But the question is not yet precise enough: in what way would history be different so that there is no moon? Are we to imagine the moon simply popping out of existence now, or the particles which formed the moon failing to coalesce and forming a ring instead, or …? The moon formed from debris spewed out from the Earth after a collision with a large planetesimal. So Comins frames one of his counterfactuals this way:

The planetesimal that created the moon traveled trillions of miles over millions of years before hitting Earth. It also swept by other planetesimals as well as by the planets Mars and Venus. Its orbit was altered by the gravitational force from each body it encountered. As a result of all these variations in its path, the planetesimal finally ended up striking the earth. But it need not have met this fate.

If that planetesimal had formed in an orbit different from its actual path by only a few inches, it would not have struck the earth. Over the planetesimal's lifetime the difference between the true orbit and any other path would have been amplified by the gravitational attractions it experienced passing near other bodies. This amplification effect, discovered in the 1980s, stems from a branch of mathematics called chaos. Had it begun in a slightly altered orbit, the planetesimal would easily have been twenty-five thousand miles to one side of its true path by the time it reached the earth in that last, fateful orbit. That change, absolutely minuscule in astronomical terms, would have prevented the collision.

Even such a near miss between earth and another body is no minor event. As it passed by, the planetesimal would be whipped into a dramatically different orbit by the gravity of the nearby earth. Depending on its new course, the planetesimal might eventually strike the sun, Jupiter, or another body, or leave the solar system forever. (ibid. 6)

The treatment of the counterfactual conditional in this passage is transparent. We are to consider a solar system just like ours (with respect to the positions and velocities of matter) save for the placement of a single planetesimal at the time of its formation. The laws of gravity are now used to determine how things would have evolved, in particular they are used to show how the difference in position would be amplified through time, enough to avoid collision with the Earth. The counterfactual has not been precisely enough specified to determine a unique further trajectory for the planetesimal, as different fates befall it in different models that meet the stated conditions.

There is more to be said about this treatment of counterfactuals (and I have tried to say some of it in my 'A Modest Proposal on Laws, Counterfactuals, and Explanations' (Chapter 1, this volume)), but it is relatively clear that once one has the laws (such as that of gravitation) in hand, the treatment of physical possibility and of counterfactuals is relatively straightforward.

If one provides a counterfactual analysis of causation (à la Lewis 1986a, chapter 21) then causal claims supervene on the counterfactuals, which are in turn underwritten by the laws of nature.

Chance also appears in physics, in two guises. There are chances that derive from stochastic dynamical equations, as in quantum theory with wave collapse. These chances are to be found written into the fundamental dynamics themselves. Then there are chances that are sometimes associated with deterministic systems, such as Buffon's calculation of the chance that a thrown needle will fall so as to intersect one of a set of parallel lines. These chances derive from a presumably natural probability measure over the possible initial conditions for the set-up. What makes such a measure natural is a somewhat vexed question, but physicists certainly take the matter to be determined (so far as it is) by other physical facts: there is never any thought of two possible worlds which agree in all their laws and in the total physical state, but disagree nonetheless on which measure over initial conditions is natural. In some cases the considerations are straightforward. Buffon, for example, uses a measure which is isotropic and homogeneous because space itself, and the various factors which influence the needle, are posited as isotropic and homogeneous. For example, if the needle were magnetized, and the lines ruled in a north–south direction, the calculation would be incorrect. In other cases, such as statistical thermodynamics, justification of a natural measure is much more subtle and difficult. But what is clear in these deterministic cases is that the chances are only as objective as the naturalness of the measure, and that in turn must be defended on other physical grounds.

So given the total physical state of the world and the laws of nature, it looks promising, and demonstrably in accord with actual scientific practice, to regard physical possibility, counterfactuals, causal connections, and chances to be fixed, insofar as they are objective at all. But what of the laws themselves?

Explicating these other notions in terms of the laws and physical state is not sufficient for Physical Statism: the laws in turn must be shown to supervene on the total physical state of the world. Is there anything in the practice of physics, classical or contemporary, which suggests that the laws themselves are determined by the total physical state?

In short, the answer is no. It matters not whether one starts with Newton, who, in the *Principia*, simply announces his three laws of motion after giving the definitions of various terms, or whether one turns directly to any contemporary textbook on quantum theory, which will treat, e.g., the Schrödinger equation as a fundamental dynamical principle. Physicists seek laws, announce laws, and use laws, but they do not even attempt to analyze them in terms of the total physical state of the universe or anything else. (One may, of course, attempt to explicate one law as a consequence or approximate consequence of another, as when showing Kepler's laws to be approximate consequences of Newton's laws of motion and gravitation in a particular situation, but that is not an attempt to analyze lawhood *per se*.) Unlike reductive analyses of possibility, causality, and chance, reductive analyses of law are not endorsed by scientific practice.

Indeed, scientific practice seems to preclude such an analysis. As we have seen, physical possibility is easily understood in terms of the models of the laws of physics. Let us suppose (and how can one deny it) that every model of a set of laws is a possible way for *a world governed by those laws* to be. Then we can ask: can two different sets of laws have models with the same physical state? Indeed they can. Minkowski space-time, the space-time of Special Relativity, is a model of the field equations of General Relativity (in particular, it is a vacuum solution). So an empty Minkowski space-time is one way the world could be if it is governed by the laws of General Relativity. But is Minkowski space-time a model *only* of the General Relativistic laws? Of course not! One could, for example, postulate that Special Relativity is the complete and accurate account of space-time structure, and produce another theory of gravitation, which would still have the vacuum Minkowski space-time as a model. So under the assumption that no possible world can be governed both by the laws of General Relativity and by a rival theory of gravity, the total physical state of the world cannot always determine the laws. The only way out is either to assert that empty Minkowski space-time must be governed by *both* sets of laws, since it is a model of both, or (a more likely move) that it can be governed by *neither* set of laws, since neither is the simplest account of space-time structure adequate to the model (the simplest account is just Special Relativity). But how can one maintain that the General Relativistic laws cannot obtain in a world that is a model of those laws, and hence allowed by them? The necessity of distinguishing the physical possibilities (i.e. the ways the world could be given that a set of laws obtains in that world) from the models of the laws signals a momentous shift from philosophical analyses that follow scientific practice to analyses that dictate it.

The situation is even worse for probabilistic laws. Consider a law that assigns a probability to any given event, say the decay of a radioactive atom. The models of such a law will include worlds where every decay event assigned a non-zero probability occurs. Hence that set of models will be *identical*, with respect to the non-probabilistic facts, to the models of a law that assigns a different probability to the event. A law that assigns radium a half-life of thirty years does not rule out every atom of radium decaying in fifteen years, it simply apportions such an eventuality a very low probability. Again, since different laws share the same models, either the laws cannot supervene on the matters of particular fact or else some models of the laws cannot be regarded as physical possibilities relative to those laws. But this last option leads to worse problems. Given a particular initial state, a probabilistic law allows many possible eventualities, and assigns each a probability (let us assume finite models). The sum of the probabilities of the various models is unity: consider, for example, a sequence of a thousand flips of a fair coin. If not all of these models are physical possibilities relative to the law, i.e. worlds where the law can hold, then a law will assign a non-zero probability to its own failure, and the sum of the probabilities of the evaluations consistent with the law will not be unity. These are indigestible consequences.

The supervenience of law on physical state, then, is not only not assumed in scientific practice, it runs contrary to that practice. We should need powerful reasons to pursue such a philosophical analysis of laws. Since Lewis does not provide one, we must seek elsewhere.

4. WHY WAS HUME HUMEAN?

The natural place to begin a search for motivations for Humeanism is Hume. Hume's reasons are quite clear, and also completely outdated, so they need little detain us.

Hume believed that every simple idea had to have been copied from a simple impression, either of perception or reflection. This raised a problem for the notion of cause and effect, since that concept included the idea of necessary connection, and the necessity of any connection between empirical events is not itself accessible to observation. Insofar as causation is analyzed by nomic subsumption, this raises a parallel problem for laws. Hume's solution was twofold: first to trace the original of the idea of necessary connection to an impression of reflection that accompanies the transition of the mind from one idea to another arising from habit, second to reduce the objective (mind-independent) conditions for causal connection to patterns of succession among event types. Hume did not see how the very idea of a cause, or a law, could ever arise in the mind if it were not somehow reducible to perceptible events. There is little need to delve more deeply into Hume's motivations, since the empiricist theory of ideas is no longer defended anywhere.

The empiricist account of concepts did not go easily. It arose again in the Logical Empiricism of the early half of the twentieth century, albeit transposed from an analysis of the *content of ideas* to an analysis of the *truth conditions of sentences*. In short, Carnap is just Hume warmed over and updated. Compare the following passages, the first from Carnap's 'The Elimination of Metaphysics through Logical Analysis of Language':

(Meaningful) statements are divided into the following kinds. First there are statements which are true solely by virtue of their logical form… They say nothing about reality. The formulae of logic and mathematics are of this kind. They are not themselves factual statements, but serve for the transformation of such statements. Secondly there are the negations of such statements ('*contradictions*'). They are self-contradictory, and hence false by virtue of their form. With respect to all other statements the decision about truth or falsehood lies in the protocol sentences. They are therefore (true or false) *empirical statements* and belong to the domain of empirical science. Any statement one desires to construct which does not fall within one of these categories becomes automatically meaningless. Since metaphysics does not want to assert analytic propositions, nor to fall within the domain of empirical science, it is compelled to employ words for which no criteria of application are specified and which are therefore devoid of sense, or else to combine meaningful words in such a way that neither an analytic (or contradictory) statement nor an empirical statement is produced. In either case pseudo-statements are the inevitable product.

Logical analysis, then pronounces the verdict of meaninglessness on any alleged knowledge that pretends to reach above or behind experience. This verdict hits, in the first place, any speculative metaphysics, any alleged knowledge by *pure thinking* or by *pure intuition* that pretends to be able to do without experience. But the verdict equally applies to the kind of metaphysics which, starting from experience, wants to acquire knowledge about that which *transcends experience* by means of special *inferences* (e.g. the neo-vitalist thesis of the directive presence of an 'entelechy' in organic processes, which supposedly cannot be understood in terms of physics; the question of the 'essence of causality', transcending the ascertainment of certain regularities of succession; the talk of the 'thing in itself') ...

Finally, the verdict of meaninglessness also hits those metaphysical movements which are usually called, improperly, epistemological movements, that is *realism* (insofar as it claims to say more than the empirical fact that the sequence of events exhibits a certain regularity, which makes the application of the inductive method possible) and its opponents: subjective *idealism*, solipsism, phenomenalism, and *positivism* (in the earlier sense). (Carnap 1959, pp. 76–7)

The second, more familiar, from Hume's *Inquiry*:

All the objects of human reason and inquiry may naturally be divided into two kinds, to wit, 'Relations of Ideas' and 'Matters of Fact'. Of the first kind are the sciences of Geometry, Algebra, and Arithmetic, and, in short, every affirmation which is either intuitively or demonstratively certain ...

Matters of fact, which are the second objects of human reason, are not ascertained in the same manner, nor is our evidence of their truth, however great, of a like nature with the foregoing. The contrary of every matter of fact is still possible, because it can never imply a contradiction and is conceived by the mind with the same facility and distinctness as if ever so conformable to reality ...

When we run over our libraries, persuaded of these principles, what havoc must we make? If we take in hand any volume—of divinity or school metaphysics for instance—let us ask, *Does it contain any abstract reasoning concerning quantity or number?* No. *Does it contain any experimental reasoning concerning matter of fact and existence?* No. Commit it then to the flames, for it can contain nothing but sophistry and illusion.

Aside from the fact that Hume is more succinct and elegant a stylist, the doctrines are nearly identical. Just like Hume, the positivists based their justification for the supervenience of law on patterns of observable events on a *semantic* thesis: any non-analytic claims that go beyond what can be empirically justified are meaningless. If one accepts this constraint, then the notion of law used by the physicists is indeed in trouble: since there can be observationally identical models of different sets of laws, the claim that a certain law obtains must go beyond what can be observed. The positivists had either to rework the notion of a law or abandon it altogether.

But no one is a positivist any more, and the shortcomings of verificationist theories of meaning are so well known as not to bear repeating. It is odd, then, that contemporary philosophers should flock to the banner of Hume.

Lewis does not announce himself a positivist, and presumably would be embarrassed at the association. The semantic theory that underlies Hume's own views has been thoroughly discredited. Why should one have 'Humean scruples' any more?

5. OTHER POSSIBILITIES

Justifications for Humean Supervenience can be divided into four categories: semantic, epistemological, methodological, and prejudicial. Semantic considerations, as we have just seen, tie the very meanings or truth conditions of sentences to matters of observable fact in such a way that any attempt to even make a claim that goes beyond those matters of fact becomes mere *flatus vocis*. Epistemological arguments can grant the *meaning* of claims that do not supervene on matters of particular observable fact, but still insist that such claims can never be *known*, or perhaps even *reasonably believed*, and so are practically idle. Methodological claims invoke some widely accepted methodological principle that implies a preference for Humean over non-Humean theories, usually on the basis of some sort of parsimony. Finally, the prejudicial stance simply declares facts that go beyond the totality of local, particular, and non-nomic facts (plus space-time) to be weird or spooky or strange. These categories are obviously somewhat arbitrary, and there is much overlap between them. The positivists, for example, consciously conflated matters of semantics and epistemology. But these four categories provide a reasonable framework for our enquiry.

Lewis, as we have seen, pegs his defense of Humean Supervenience to materialism, i.e. physicalism, but that provides no motivation for the reduction of laws to something else. Hume's and Carnap's semantic views are thin reeds, to say the least. Here is what John Earman has to say when the question of justification comes up (see quotation cited above):

The well-known motivations for (E1) fall into two related categories. There is ontological argument, intuition, and sloganeering ('The world is a world of occurrent facts'), the three often being hard to distinguish. Then there are epistemological arguments and threatenings, the most widely used being the threat of unknowability, based on two premises: we can in principle know directly or non-inferentially only (some subset of) occurrent fact, what is underdetermined by everything we can know in principle is unknowable in principle .... The argument connects back to the ontological if we add the further premise that what isn't knowable in principle isn't in principle. (Earman 1984, p. 195)

Earman's characterization is telling in at least three ways. The first is that Earman himself, although sympathetic to the empiricist position, does not directly endorse any of these motivations. Second is the introduction of the term 'occurrent' into the discussion (but see also how Earman finesses the question of the meaning of 'occurrent' in the passage cited earlier). If only local matters of particular non-nomic fact, and logical combinations of them, are 'occurrent', then accepting anything that goes beyond these is accepting something non-occurrent. But 'non-occurrent' sounds suspiciously like 'not really happening', or perhaps 'not really there'. The exact meaning of 'occurrent', and the appropriate conditions for its use, deserve more notice.

The standard means of explicating the notion of an occurrent property is to contrast it with a *merely* dispositional one, fragility being the usual example. It is true that the window is fragile, and true that it is massive, but the ontological status of massiveness and fragility are quite different. Massiveness is just a matter of how the window *is*, while fragility is a matter of *how it would behave under certain circumstances.* One then asserts that in order to be legitimate, the truth conditions for claims about dispositional properties must ultimately reduce to matters of occurrent fact: there are no 'free floating' dispositions. This line becomes a bit embarrassing if dispositional accounts of apparently occurrent properties are offered (e.g. if having a mass is analyzed in terms of how an object would behave if subject to a force), but let's leave this wrinkle aside.

Insofar as 'non-occurrent' carries the implication 'not really there in its own right', then the use of it *at the beginning of an ontological discussion* is obviously question-begging. I believe that there are laws at every point of spacetime, and that the laws cannot be reduced to facts about the total physical state of the universe. In some straightforward sense, I therefore believe that the laws are occurrent. If someone replies that on their understanding of 'occurrent', laws just aren't the sorts of thing that *could* be occurrent, then I would respond that on their understanding of 'occurrent', the inference from 'non-occurrent' to 'in need of reduction' is invalid.

What is the real problem with fragility? Fragility is not as 'real' a property as mass for the simple reason that we take it to supervene on fundamental physical and chemical properties. Once you have specified the exact physical composition of the window, *and given the physical laws*, it follows by analysis that the window will break if a rock is thrown at it with sufficient force. Fragility is not a *fundamental* physical property, in that two pieces of glass cannot be physically identical save for their fragility: if one is fragile and the other isn't, then there must be some other difference between them at the level of their physical composition. Mass and charge *are* fundamental physical properties in that particles *can* differ solely in mass (e.g. electrons and muons) or solely in charge (e.g. electrons and positrons) without there being any further physical difference which accounts for this. An ontology that accepts fragility at a fundamental level, alongside mass and charge, is both too bloated and in danger of contradiction. Too bloated because all of the behavior of the window can be predicted and explained (we believe) from its description in terms of mass, charge, etc. In danger because it suggests that two windows could agree in all other physical respects but differ in fragility, which we take to be false. It is a further notable fact that fragility, unlike mass and charge, does not figure in the fundamental laws of physics.

The reduction of fragility, though, is of necessity a reduction to both fundamental physical state *and law*. And none of the reasons for believing that fragility can be explicated in terms of physical state and law can be used to argue that law itself can be explicated in terms of physical state alone. As we have seen, while science is in the business of explaining things like the fragility of a piece of glass, it is not in the business of giving reductive analyses of the laws of nature.

The third revealing aspect of Earman's survey of motivations is the introduction of epistemology, together with the qualifier 'in principle'. On the assumption that all atomic observations are observations of matters of local, particular, non-nomic fact and their space-time relations, the non-supervenience of laws on the total physical state of the universe implies that there could be two universes that are observationally identical, yet differ with respect to their laws. What it would 'feel like' to live in either of these universes would be exactly the same, yet the facts which obtain in them would differ. Hard positivists would deny that any such seeming difference could be real, or the words used to describe the apparent differences meaningful, but we have let hard positivism go its way. Still, there is something slightly spooky about facts that go beyond the evidence. Let's see if we can exorcize this ghost.

Let us first recognize that the existence of facts which are not determined by the complete totality of all observations, past, present, and future, is commonplace. No doubt, Socrates had a blood type. Also, no doubt, one could not deduce that blood type from a complete catalogue of every observation that has been or will be made. No test of the requisite type was ever made on Socrates, and, doubtless, no remains that could be identified as his will be subject to such a test. Socrates' blood type is now, and will always be, beyond our ken.

Of course, only a lunatic would conclude that the ontological status of Socrates' blood is thereby affected, that he (miraculously) had no blood type at all, or an indeterminate one. No doubt there was a fact of the matter about his blood, despite our irremediable inability to know it. So the question arises: what difference does it make to ontology that a particular fact cannot be deduced from actual evidence?

This question is likely to strike one as trifling. After all, as Earman's quote indicates, the issue should not be what is, or was, or will be known, but rather *what is knowable in principle*. Socrates never had his blood tested but he *could* have had it tested (in some sense of 'could'), and the test *would have* revealed the blood type. Socrates' blood type in not ontologically worrisome because it was knowable in principle.

This sort of response is so common and well entrenched that it takes an effort to see how utterly bizarre it is. There are straightforward epistemological problems about Socrates' blood type: we don't know what it was and we never will. The question is whether this epistemological problem should have any implications for ontology at all. The commonsense answer is no. And the philosopher wishing to maintain the commonsense response, but still desiring to link ontology to epistemology, makes the link not to *actual* evidence but to *merely possible* evidence. But if I am somehow worried about the ontological status of Socrates' blood type for epistemological reasons, how will the invocation of *counterfactual assertions about merely possible observations* allay those worries? Is anyone really supposed to say: 'Well, I was initially concerned about how "Socrates had blood type O" could have a determinate truth value, but my fears have been allayed by giving truth conditions in terms of counterfactuals, for I have no qualms about accepting that the counterfactuals have definite (but unknown) truth values'!? This has everything exactly backwards: we think that there is a determinate (but unknown) fact about how such tests would have come out exactly because we think there is a determinate (but unknown) fact about what the blood type was, and that the testing procedures would have revealed it. Relying on the counterfactuals to somehow validate the use of plain indicatives ('Socrates' blood type was O') is both baroque and self-defeating.

On any view, the laws of nature are not knowable *in fact* in the sense that they do not follow deductively from all the observations there have been, are, and will be. On my telling, they are not knowable even *in principle* in that they do not follow deductively from everything in the history of the universe that *could* have been observed. What follows? If severe ontological worries follow unknowability in principle, it is hard to see why the same worries won't accompany unknowability in fact, and that fate strikes most of the events in the history of the universe. Indeed, knowability in fact is already a philosopher's fantasy: there is no catalogue of observations being recorded in the epistemologist's book of life, such that all actual observations are accessible to us. The practical problem of epistemology is inference from data we have to hand, and from this essentially all of the universe, including the recent past and all of the future, is underdetermined. If ontology follows epistemology this far, there will be precious little of existence left to us at all.

There is a milder form of epistemological upset than unknowability in principle. This form admits that we cannot be constrained in our ontology only to those things that follow from our observations, so ampliative inferences must be allowed. But still, we should distinguish those ampliative inferences that can be (in some instances) checked from those which cannot. Induction, for example, is always a leap beyond the known, but we are constantly assured by later experience that we have landed safely. Inference from observations to laws of nature, on the other hand, can never be later vindicated since all the data give us are patterns of particular matters of fact.

This is a serious concern, and a complete answer to it is beyond our present scope. Suffice it to say that this sort of skeptical worry about ampliative inferences also strikes theoretical entities such as quarks, which are never directly observed. Suspending belief about unobservable entities is a time-worn strategy, albeit one uniformly rejected by actual scientific practice. And if one is agnostic about things like electrons and quarks, not to mention quark color and flavors, then the question of laws of nature, as understood in scientific practice, is already decided, since the laws of fundamental physics are couched in those terms. But this is not the sort of view that a scientific realist such as Lewis would endorse.

Our epistemological access to most matters of particular, local, non-nomic fact already demands ampliative inferences that can never be fully vindicated by experience. That is why an *epistemological* defense of Humean Supervenience cannot be satisfied merely with a reduction of laws to total physical state, or total local physical state, but must insist on *total observed physical state*. But this so handcuffs ontology that few would be willing to accept it. Thus, as cited above, Earman writes:

> There may also be general occurrent facts (I think there are), but these are also parasitic on the singular occurrent facts. Conservative empiricists may want to

restrict the antecedent of (E1) so as to range only over observable facts while more liberal empiricists may be happy with unobservable facts such as the fact that quark q is charming and flavorful at t. In this way we arrive at many different versions of the loyalty test, one for each persuasion of empiricist. (ibid.)

Lewis would clearly be a liberal empiricist on this telling, but just what sort of an empiricist is a liberal empiricist at all? Quarks and their flavors, as well as wavefunctions, are neither the data of experience nor constructs from them. Of course, we only believe in quarks and wavefunctions on the basis of observations, not a priori, but the same is true of laws of nature, such as the relativistic field equations. Lewis's penchant for the singular, local, and non-nomic cannot be given an epistemological foundation, for attempts to appeal to epistemology will banish a lot more than non-supervenient laws.

There is one final methodological consideration that might be used to defend Lewis's version of Humean Supervenience: Ockham's Razor. Any physicalist already admits the total physical state of the world into his or her ontology, and thereby also anything that supervenes on the total physical state. On the general principle that less is more, why not try to live as economically as possible, preferring not to go beyond the total physical state unless pushed?

The question is how much of a push is needed. While the Razor is demonstrably good methodological advice in some circumstances (in particular, when it councils higher credence to explanations which posit a single cause to multiple events that occur in a striking pattern over explanations invoking coincidental multiple causes), it is hardly universally accepted. Very few physicists, for example, still pursue pure relationist theories which eschew space-time for spatio-temporal relations between material objects. The relationist ontology is strictly a subset of the usual physical ontology, since those who countenance space-time also accept that there are material objects with spatio-temporal relations, they simply do not think that to be all there is.[6] But for all that, physicists are not rushing to replace field theory with action-at-a-distance particle theories, in hopes of reducing ontology. The relationist theories are just too complicated and contrived.

---

[6] One can argue for an understanding of spatio-temporal structure in which there are no spatio-temporal relations, if by 'relations' one means something whose existence only requires the existence of the relata. On this understanding, no spatio-temporal relations can be 'skimmed off' the substantivalist ontology. See 'Suggestions from Physics for Deep Metaphysics', Chapter 3, this volume.

Similarly, one could try to live within the confines of the total physical state of the world, and find good enough substitutes for laws that can be defined in those terms, but why do so? One would take fewer chances thereby (though barely fewer, since the total physical state is itself largely unknown), but at what price? One could also refuse to assert or deny counterfactual claims in general, and thereby take fewer risks of being wrong, but it seems like a poor and miserly existence.

Lewis's own stated motivation is robust and healthy. Ontology is the general account of what there is, and our knowledge of what there is is grounded in empirical science, not in a priori speculation or prejudice. Philosophical accounts that force upon us something that physics rejects ought to be viewed with suspicion. But equally suspect are philosophical scruples that rule out what physics happily acknowledges. As we have seen, contemporary physics posits physical facts that are Non-Separable. What grounding could a preference for Separability have to suggest that we ought to warp either the physics itself, or our account of space to accommodate Separability? And physics has always postulated that there are laws without suggesting that they supervene on or reduce to matters of particular non-nomic fact. Hume, armed with an empiricist semantics, had reason (by his own lights) to be worried. But we no longer accept Hume's account of concept formation and its allied account of linguistic meaning and truth. So the question remains: why be Humean?

3

Suggestions from Physics for Deep Metaphysics

Begin with two theses about metaphysics, one foundational and the other programmatic.

First: metaphysics, i.e. ontology, is the most generic account of what exists, and since our knowledge of what exists in the physical world rests on empirical evidence, metaphysics must be informed by empirical science. As simple and transparent as this claim seems, it would be difficult to overestimate its significance for metaphysics. Kant, for example, maintained that metaphysics must be a body of necessary truths, and that necessary truths must be a priori, so metaphysical claims could not be justified by experience. The logical empiricists followed Kant's lead on the epistemological standing of metaphysics, and so rejected the entire enterprise out of hand. The Kantian strain survives yet in the notion that the proper object of metaphysical study is not the world, but rather our own 'conceptual system', particularly insofar as that conceptual system structures our thought about the world. This tack allows metaphysics to remain a priori (on the assumption that we can discover facts about our conceptual system merely by reflection), while depriving it of any pretension to be about the world itself. Any attendant sense of loss may be mitigated by insisting that talk about 'the world itself, as it is independently of our thought' is itself meaningless or confused.

Modern physics, i.e. Relativity and quantum theory, annihilates Kant's theses about the status of space and time and causality: whatever one may think about the world revealed by experience, one cannot think that it must be presented to us as in Newtonian space and time, governed by deterministic laws. Empirical science has produced more astonishing suggestions about

---

I would like to thank the departments of Philosophy at the University of North Carolina, Chapel Hill, CUNY, and MIT, for providing wonderful venues and spirited discussion of the early incarnations of this chapter. I am especially indebted to Geoff Sayre-McCord for sound advice regarding the exposition.

the fundamental structure of the world than philosophers have been able to invent, and we must attend to those suggestions. That our physical theories are supported by empirical evidence is no demerit, but rather provides us with grounds for believing that these extravagant accounts of what exists might be correct.

Second: having rejected the notion that we cannot possibly escape the confines of our present 'conceptual system' to discover how the world is, one naturally begins to look with suspicion upon our system of representations rather as *impediments* to understanding the world. More particularly, it seems advisable to guard against the error of mistaking the structure of language for the structure of the world itself. This has been nicely expressed by Bertrand Russell:

> Reflection on philosophical problems has convinced me that a much larger number than I used to think, or than is generally thought, are connected with the principles of symbolism, that is to say, with the relation between what means and what is meant. In dealing with highly abstract matters it is much easier to grasp the symbols (usually words) than it is to grasp what they stand for. The result of this is that almost all thinking that purports to be philosophical or logical consists in attributing to the world properties of language. (Russell 1923, p. 84)

If Russell is right, we commonly interpret features of a representation as features of the thing represented, thereby illegitimately projecting the structure of our language onto the world. The most commonly diagnosed form of this malady is hypostatization—the belief that an entity must correspond to a given nominal linguistic form—but there are any number of ways in which our accounts of what exists could be influenced by the grammatical form of our language or the structure of a mathematical representation.

One of the most suggestive examples of the parallelism between grammar and ontology is provided by Aristotle. In the *Categories*, Aristotle distinguishes all entities into various ultimate genera, giving pride of place to the genus *substance*. And one of the primary marks of substance, according to Aristotle, is being the ultimate subject of predication: items from the categories other than substance only exist at all insofar as they are predicated of substances (and the so-called secondary substances only exist in virtue of being predicated of the primary substances). Aristotle's views are quite difficult to interpret, and the *Categories* is made even more opaque by unclarity about whether the relata of the predication relation are supposed to be *words* or *things*. Indeed, this very unclarity suggests that Aristotle may have fallen prey to Russell's error, and have mistaken grammatical form for ontological structure. More particularly, it is a possibility worthy of consideration that Aristotle has been led to his metaphysics of *substance* and *universal* by projecting out the linguistic forms of *subject* and *predicate*.[1]

Universals have been regarded with a jaundiced eye throughout philosophical history. And the nominalist dictum—items described by the same term have nothing in common but the name—suggests exactly that the ontological doctrine of universals is founded on a naive understanding of the significance of linguistic form. But it is one thing to cast doubt on the existence of universals, and something else to describe an adequate alternative to the doctrine. I think that the outline of such an alternative exists, having been discovered not by philosophers but by physicists, under the rubric *gauge theory*. The primary object of this paper is to sketch enough of that theory to make clear how it constitutes an alternative to the theory of universals, and to the theory of tropes, and to all of the various kindred metaphysical hypotheses that are considered in the philosophical literature. We will make our way slowly to gauge theory, beginning with a brief and biased account of one influential approach to ontological questions that derives from the work of W. V. O. Quine.

1. QUINE AND MODERN ANALYTICAL METAPHYSICS

In 'On What There Is' (1953), Quine offered a very attractive account of ontological commitment. The slogan is 'to be is to be the value of a variable', but the slogan taken at face value is misleading. Quine is not suggesting that existence itself somehow depends on variables—otherwise nothing would exist without language, and one would rightly wonder how language could have come into existence. Quine's most immediate doctrine is not about existence itself, but about ontological commitment: what does a theory *say* there is in the world? The first pass answer is: a theory is ontologically committed only to whatever sorts of entities would have to be in the domain of the variables of the theory if the theory is to be true. So to determine what a theory says there is, translate it into a formal language and see what sorts of models could make the theory true. This alone solves the question Quine starts out with: how can one meaningfully deny the existence of an object without thereby being implicitly committed to its existence?

---

[1] For our purposes, the 'metaphysics of substance and universal' is neutral between the Aristotelian doctrine of *universalia in rebus* and the Platonic doctrine of *universalia ante rem*.

Suggestions for Deep Metaphysics

Insofar as Quine's topic is merely ontological commitment, the title of the paper ought to be 'On What a Theory Says There Is'. But it is easy enough to see how the doctrine of ontological commitment can become a central element in a theory of ontology: one merely adds that a particular theory is, in fact, true. If the theory is true, and if Quine has correctly explained how to determine what must exist in order that the theory be true, then we can tell something about what there is. This suggests a three-step procedure for approaching ontological questions, which we may call *Quine's Recipe*:

Step 1: Settle on some sentences that you take to be true.

Step 2: Translate these sentences into a formalized language, such as first-order (or second-order) predicate calculus.

Step 3: Determine what must be in the domain of quantification of the variables of the translated sentences in order that they come out true.

Carnap had already used the Recipe to show why accepting 'Rain is outside' commits one to the existence of rain, while accepting 'Nothing is outside' does not commit one to the existence of Nothing[2], and Quine shows how following the Recipe allows us to accept the truth of 'Pegasus does not exist' without somehow being forced thereby to accept the existence of Pegasus. It is easy to see how David Lewis's arguments for the existence of possible worlds fit this mould as well.

There are some features of Quine's Recipe that deserve our close attention. One is the importance of the second step, translation into a formalized language, in the scheme. Sentences with the same surface grammar in natural language could end up having very different sorts of ontological implications if they are translated differently, as the example of 'Rain is outside'/'Nothing is outside' already shows. Let's illustrate this with some examples. Consider the English sentences:

1: Clio and Maxwell have the same father
2: 'cold-hearted' and 'unsympathetic' have the same meaning.

The surface grammar in English of these sentences is identical, but Quine's Recipe allows us to maintain that they have different sorts of ontological implications if we just provide them with different sorts of translations. The natural translation of 1 into first-order predicate calculus is:

$$1': (\exists x)(Fxc \& Fxm)$$

[2] In Carnap 1959.

If we add the further postulate that no one is his or her own father ($\sim (\exists x)(Fxx)$), then the truth of 1' would require the existence of at least one entity beside Clio and Maxwell, as intuitively we think it should.

Mechanically applying the same analysis to 2, though, yields the translation:

$$2' (\exists x)(\text{MEAN}x\text{'cold-hearted'} \ \& \ \text{MEAN}x\text{'unsympathetic'})$$

where 'MEANxy' denotes the relation between a meaning and a word which has that meaning. So apparently accepting the truth of 2 implicitly commits us to something else in our ontology beside words: namely meanings that the words can 'have'. But Quine himself (in 'Two Dogmas of Empiricism', 1951) rejects exactly this inference. There he says that there is nothing to prevent us from understanding the truth of sentences like 2 in terms of a relation of *synonymy* between words, so the proper translation of 2 is just

$$2'' : \text{SYN'cold-hearted''unsympathetic'},$$

a two-place relation that holds directly between the words. By translating 2 as 2'' rather than 2', we eliminate our commitment to meanings as entities, a result that warms the heart of Quine *qua* lover-of-desert-landscapes.

But it must also be noted that naive translation from natural language into formal language can obscure necessary ontology as easily as it can create superfluous ontological commitment. The surface grammar of

3: Clio and Maxwell are siblings,

suggests the translation

$$3' : \text{SIBcm},$$

which would imply that the truth of 3 has as its ontological commitment only the existence of Clio and Maxwell. But this is incorrect: if no one can be his or her own parent, then the truth of 3 requires that more than Clio and Maxwell exist in the world. The 'deep ontological structure' of 3 is expressed by

$$3'' : (\exists x)(Fxc \ \& \ Fxm) \ \& \ (\exists x)(Mxc \ \& \ Mxm),$$

where 'Mxy' represents the motherhood relation. Given the principles that no one is his or her own mother or father, and no one is both the mother and father of anyone, i.e. adding to our theory

$$\sim (\exists x)(Fxx)$$
$$\sim (\exists x)(Mxx)$$

and

$$\sim (\exists x)(\exists y)[(Fxy) \,\&\, (Mxy)],$$

we find that the truth of 3 demands the existence of at least two items beside Clio and Maxwell. The point is yet more obvious for, say, second cousins: even God could not have created Adam and Eve as second cousins.

So the dictum 'To be is to be the value of a variable' does not give us as much guidance as we might have hoped to deep questions of ontological commitment. Once the translation into formal language has been made, the dictum comes into play, but the lion's share of real metaphysical work is done when settling on the right translation. Quine himself does not give us much guidance on this point, but it is clear that deciding on which translation to use is part and parcel of deciding on the most acceptable theory about the deep structure of certain properties and relations. As we will see, considerations such as explanatory power enter into these sorts of decisions.

Quine was more concerned in 'On What There Is' to defend his brand of nominalism than to investigate exactly how one determines whether the truth of a sentence like 3 demands the existence of two entities or four. His immediate target was universals, and he considered it enough to have shown that accepting the truth of 'This house is red' and 'That sunset is red' does not commit one to the existence of anything save the house and the sunset. In particular, it need not commit one to a universal 'Red' or 'Redness' which both the house and the sunset *have* or *exemplify*. One *could* become ontologically committed to some such entity, but only by employing second-order logic as the formal language used in translation. Following this line of thought, the question of whether universals exist seems to reduce to the question of the utility, or necessity, of using second-order rather then first-order logic.

But there is something unsatisfactory about this facet of Quine's approach to ontology. If one restricts oneself to first-order predicate calculus, then it seems that the only *ontological* implications of a theory concern the *subjects* of which predicates are predicated. The nature of the predicates, and what conditions must hold in order for the predicate to be *true of* a subject, seems to have disappeared from the radar screen. If red houses and red sunsets have nothing in common, why is it true to call them both *red* and, more directly, why is it that we are even *inclined* to apply the predicate 'red' to both of them? This is the question which medieval nominalism appears impotent to answer: if red houses and red sunsets have nothing in common but the name 'red', what could account for the fact that they have the name in common in the first place? Must there not be something about the *nature* of the two objects that explains our linguistic behavior towards them?

If one ignores the significance of the predicates in a language entirely, then the question 'what must the domain of quantification of a theory be like for the theory to be true?' reduces to a matter of nothing but the *cardinality* of the domain. The purely logical structure of a first-order theory can place constraints on the sort of universe that can serve as a model of the theory, but the restriction amounts to no more than a restriction on the *number of objects* in the domain. Down this road ultimately lie Putnam's model-theoretic arguments against 'metaphysical realism'. Down this road (albeit via a different fork) also lies Quinean nominalism: the view that all one needs in one's ontology is a collection of first-order objects and all of the sets constructible from them. For once one has the sets, one can associate monadic predicates with sets of objects (their extensions), relations with sets of sets, and so on.

Set-theoretic extensionalism of this sort appears to become even more powerful when supplemented by David Lewis's modal realism (surely an unholy alliance in Quine's eyes): given transworld extensions, 'creature with a heart' can be distinguished from 'creature with a kidney', propositions can be identified with sets of worlds, etc. But, as Lewis recognized, pure extensionalism cannot suffice. Even if one 'identifies' properties with sets of objects, it cannot be that every set of objects constitutes an equally good property, in all senses of 'property'. If it did, then any pair of objects would be as similar as any other pair, since every pair would have exactly the same number of common properties:

> There is another great rift in our talk of properties. Sometimes we conceive of properties as *abundant*, sometimes as *sparse*. The abundant properties may be as extrinsic, as gruesomely gerrymandered, as miscellaneously disjunctive, as you please. They pay no heed to the qualitative joints, but carve things up every which way. Sharing them has nothing to do with similarity. Perfect duplicates share countless properties and fail to share countless others; things as different as can be imagined do exactly the same. The abundant properties far outrun the predicates of any language we could possibly possess. There is one of them for any condition we could possibly write down, even if we could write at infinite length and even if we could name all the things that must remain nameless because they fall outside our acquaintance. In fact, the properties are as abundant as the sets themselves, because for any set whatever, there is the property of belonging to that set. It is these abundant properties, of course, that I have identified with the sets.

The sparse properties are another story. Sharing them makes for qualitative similarity, they carve at the joints, they are intrinsic, they are highly specific, the sets of their instances are *ipso facto* not entirely miscellaneous, there are only just enough of them to characterize things completely and without redundancy. (Lewis 1986b, pp. 59–60)

Lewis goes on to survey three competing accounts of the nature of sparse properties: (1) the extension of a sparse property is the set of objects which instantiate one and the same universal, (2) the extension of a sparse property is the set of objects which have as parts duplicates of some trope, (3) the extension of a sparse property is a set which is distinguished by a further unanalyzable, primitive characteristic of naturalness. 'In the contest between these three alternatives—primitive naturalness, universals, or tropes—I think the honors are roughly even, and I remain undecided' (ibid. 64).

None of Lewis's options demands that a natural property correspond to every meaningful predicate: truth conditions for less-than-natural properties could be constructed from the natural ones and logical operations, for example. This sort of project has been carried out for universals by David Armstrong in his scientific realist theory of universals (Armstrong 1978). Armstrong insists that we must let science inform us which predicates correspond to real universals, of which there are likely to be only a few. But the main point we must insist on is still Lewis's: the whole *point* of admitting either universals or tropes or natural sets is to account for the objective *similarity* and *dissimilarity* of objects. For if there are no objective facts about the comparative character of objects, we must fall back into the unpalatable position that the only real structure of the universe is its cardinality. And the problem with Quine's Recipe, as it stands, is that although it may help us identify what objects a theory is committed to, it leaves us in the dark as to the metaphysical significance of the predicates and the nature of their truth conditions.

For the remainder of this essay we will consider only the theory of universals, understanding that similar remarks could be made about the theory of tropes and of natural sets. The traditional theory of universals distinguishes universals into properties and relations, and relations further into internal and external. Whether an internal relation holds between some objects is determined by the intrinsic characteristics of the objects, characteristics each could have on its own. An external relation, in contrast, is something over and above the intrinsic character of the relata. Knowing all there is to be known of the relata taken individually will not inform one about the external relations between them. Thus we arrive at a fundamental metaphysical picture: in the world, there are objects, which are referred to by singular terms and quantified over by first-order variables, and there is some structure to the set of objects provided by universals (or tropes, or primitive naturalness) which plays a fundamental role in the explanation of why predicates have the extensions that they do. There is, of course, something suspicious in the way the subject/predicate structure of the language seems to be mirrored in the ontology, but the question remains: what is the alternative?

2. A NEW PROGRAM

The project of undermining the traditional metaphysical picture sketched above will proceed in three stages. First, I will argue that there are no metaphysically pure external relations. Next, that there are no metaphysically pure internal relations. And since internal relations are supposed to be determined by the intrinsic properties of objects, it follows that there are no metaphysically pure intrinsic properties. These arguments will be predicated on an alternative account of ontological structure provided by contemporary physics. In the end, I will qualify the most radical conclusions a bit, but the qualification still leaves the main thrust of the argument intact.

We begin by clarifying the notion of a *metaphysically pure* relation or property. Intuitively, the relation of being siblings is not metaphysically pure, since, as we have seen, for two people to be siblings the existence of more than the two people is required: there must also be the parents. Of course, if people could be their own parents, then siblinghood would not necessarily demand the existence of more than the two relata, but even then the relation would not be a metaphysically pure, direct, further unanalysable relation between the relata: siblinghood only obtains in virtue of the obtaining of other relations between things.

This exposition does not completely clear up the matter of metaphysical purity, and there are doubtless difficult cases, but fortunately all we will need in the sequel is a *necessary* condition for a relation to be metaphysically pure. If a relation is metaphysically pure, then it is at least *possible* that the relation be instantiated in a world in which only the relata of the relation exist. This is not a sufficient condition, but seems unexceptionable as a necessary condition: if it fails, then the condition for the holding of the relation must make implicit reference to items other than the relata, so the relation is not just a matter of how the relata directly stand to each other.

Suggestions for Deep Metaphysics

Similarly, if a property is metaphysically pure then it must be at least possible for the property to be instantiated in a world in which only the item that has the property exists.

The first order of business is to argue that *there are no metaphysically pure external relations*. The most uncontroversial examples of fundamental external relations are the spatio-temporal relations. Indeed, there is no uncontroversial example of an external relation that does not involve space and time. If one believed in a primitive causal relation between events, then there could be worlds which agree in their spatio-temporal structure, and in the distribution of properties in the space-time, and agree on all laws which are not defined in terms of the causal relation, but disagree about which events cause which. I see no reason to accept this as a metaphysical possibility, and so reject the notion of a fundamental causal relation.[3] Since I can think of no other plausible candidate for a non-spatio-temporal external relation, I will confine my remarks to space and time. The argument would have to be augmented if further candidates emerge.

What, then, is the metaphysical structure of the fundamental spatial relation, the distance between two things?[4] In particular, can two objects be some distance apart in a world that contains only those two things?

The answer, I claim, is 'no'. Distance between two points is a metaphysically derivative notion. The primitive notion from which it is derived is the length of a path in space. The distance between two points is then defined as the minimal length of a continuous path that connects the points. Just as the relation of siblinghood metaphysically involves a hidden quantification over people who are related as parents to the siblings, so the relation of distance between points involves a hidden quantification over all the continuous paths that connect the points. If two points cannot be connected by a continuous path, then *ipso facto* it is metaphysically impossible for there to be any distance between them.

One might reply to this argument as follows: while it is true that given the notion of the length of a path, the distance between points can be defined, it is equally true that given the notion of distance between two points, the length of a path can be defined. Suppose one is given a path from A to B. The path can be subdivided into pieces by specifying a sequence of points on it between A and B. Call the sequence $p_1, p_2, p_3, \ldots p_n$. If the distance between any pair of points is given, then among these are the distance from A to p_1, the distance

---

[3] See 'Causation, Counterfactuals, and the Third Factor', Chapter 5, this volume.
[4] For ease of exposition, I will confine my remarks to a classical picture of spatial structure. The application to relativistic space-time structure is obvious.

from p_1 to p_2, ... and the distance from p_n to B. The sum of these distances is an approximation to the length of the path. The length of the path is then defined as the limit of this sum as the path is divided into more pieces by points whose maximum distance to their adjacent points approaches zero. So if length along a path can be defined via distance between points just as well as distance between points can be defined via length along a path, by what right can we nominate the notion of path length as the more fundamental?

The answer is that distances between points must be postulated to obey certain constraints, and these constraints can be derived from the definition of distance in terms of length. In order for a set of distance relations between a trio of points to be instantiated by points in some space, they must obey the Triangle Inequality: the distance from A to B plus the distance from B to C must be at least as great as the distance from A to C. Why should this inequality hold? In other sorts of relations, no such inequality is demanded: if we could quantify degrees of mutual animosity, for example, we would not find that the animosity between A and B plus the animosity between B and C must be at least as great as the animosity between A and C. And for any metaphysically pure external relation we should expect no such constraints: since the relation is external, its holding puts no constraints on the relata, and since it is metaphysically pure, its holding between the relata puts no constraints on anything else. It seems that it ought to be metaphysically possible for a trio of entities to have any triple of metaphysically pure external relations among them.

If distance is defined in terms of paths, on the other hand, then the Triangle Inequality falls out automatically. Since a path from A to B connected to a path from B to C *is* a path from A to C, the minimal length of a path from A to C cannot be greater than the sum of the length of the minimal path from A to B and the length of the minimal path from B to C. This is why relationists about space and time (e.g. Julian Barbour in his (2000), see p. 42) have to brazenly postulate primitive constraints on the sets of distance relations that can be simultaneously instantiated between items, while substantivalists (who believe in the paths even when they are not occupied by material objects) need no such constraints.[5] And the set of constraints on distances is not exhausted by the

---

[5] This particular issue turns not on the relationism/substantivalism debate *per se*, but on whether it is metaphysically necessary for there to be a continuous path connecting any points which have a distance between them. A relationist who has principles that metaphysically guarantee a plenum could adopt this account of distance, although such a relationist becomes hard to distinguish from a substantivalist. See my (1993).

Triangle Inequality: even if the distances between every triple of points drawn from a quadruple satisfy the Triangle Inequality, it does not follow that the distances for the whole quadruple can be embedded in a space. There are separate, ever more complicated constraints for quadruples, quintuples, and so on.

So we ought to take length along a path as primitive, and define distance in terms of it. And it follows that two points which cannot be connected by a continuous path cannot have any distance between them, so all points related by distance to one another must be parts of a single, common, connected space. The existence of a distance between any pair of points therefore metaphysically requires that there be more in the world than the two points: distance is not metaphysically pure.[6]

Temporal relations work similarly, as do the relativistic spatio-temporal relations. So if spatial and temporal relations are the only fundamental external relations, then there are no metaphysically pure external relations.

Our next order of business is to argue that there are no metaphysically pure internal relations. This is where gauge theories come into consideration, but we will work our way up to them by beginning with a simple example. Consider, then, the intuitive notion of an arrow *pointing in a certain direction*, and, more specifically, the notion of two arrows at different locations *pointing in the same direction*. Having cut our teeth on Euclidean geometry, we

---

[6] The foregoing discussion assumes that space remains continuous at all scales. But what if, at some very small scale, space is discrete? One possibility is that space is made up of individual points, any pair of which either bear or do not bear an *adjacency* relation. The adjacency relation would be a metaphysically pure external relation: no further facts would determine whether two points are adjacent or not. Space could then be thought of as a lattice, with the nodes representing spatial points and the edges connecting adjacent points. The distance between adjacent points could be defined to be a unit, and the distance between non-adjacent points would be the minimal path length connecting them in the lattice.

Such a scheme would allow for the existence of a metaphysically pure external relation, the adjacency relation. But several observations are in order.

First, it would still be true that *macroscopic* distances would not be metaphysically pure: they would be defined in terms of minimal paths as just described.

Second, the adjacency relation would not itself be subject to any constraint analogous to the Triangle Inequality: it would be metaphysically possible for a triple of points to all be adjacent to one another, or none adjacent, or any one pair, or any two pairs. So the argument against the relation being fundamental would disappear. But …

Third, some sort of mysterious constraint would have to be invoked concerning the totality of adjacency relations. For if one takes, say, a million points, and arbitrarily assigns pairs of points as adjacent, and then calculates distances via path lengths as described above, then one will get a set of distances between points that satisfy the Triangle Inequality. But there is no reason at all to suspect that the resulting set of distance relations will be approximated by the distances among a million points in a *single three-dimensional space*. Only a very few such resulting lattices will yield a structure which approximates a small-dimensional continuous space. So we would still be left with the mystery of why the discrete space so nearly mimics a low-dimensional continuous space.

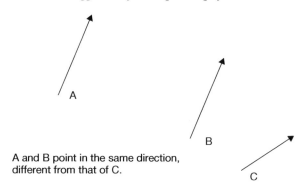

A and B point in the same direction, different from that of C.

Figure 1. Parallel Directions in Euclidean Space

naturally regard it as an absolute fact about arrows located at different spatial locations that they either point in the same direction or in different directions; they are either parallel or not (see Figure 1).

It is perhaps not so obvious whether pointing in the same direction should count as an internal or external relation between arrows, but in any case it seems intuitively to *be* a relation, to either hold or fail to hold between any pair of arrows.

A modicum of non-Euclidean geometry, though, disabuses us of these ideas. Take, for example, the two-dimensional geometry of the surface of a sphere. At any point on the surface, there are many directions that one can go in. The set of directions looks structurally the same at all points. But if we ask whether a certain arrow at a middling latitude points in the same as or different direction from an arrow at the equator, no answer is immediately apparent. On the sphere, the set of directions at any given point live in a tangent space which is attached to that point, and admit of no immediate comparison with the directions in any other tangent space (Figure 2).

From the fact that the directions at a point live in a unique tangent space, one might conclude that comparisons between directions at different points are completely impossible: each set of directions is *sui generis* and incomparable with any other set. But this conclusion would be too extreme: it is both possible and important to make certain sorts of comparisons between directions at distinct points. What we need is something that, as it were, ties the various tangent spaces together. The relevant mathematical gadget is called an *affine connection*, and in rough terms the affine connection allows one to make comparisons between the directions at one point and the

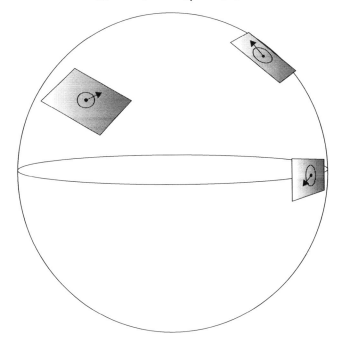

Figure 2. Directions at Different Points on a Sphere

directions at points that are infinitesimally nearby. More precisely, the affine connection allows one to define *parallel transport*: carrying an arrow from one point in the surface to another *along a continuous path* without ever 'twisting' it or changing its direction.

Our immediate intuitions about parallel transport on the surface of a sphere tend to be quite reliable. Suppose, for example, one is commanded to transport the arrow A at the North Pole p to point q at the equator *along path* α, always keeping the arrow parallel. The result will be arrow B (Figure 3).

If we then transport B from q to r along path β, the result will be C. So one might say that C 'points in the same direction' as A, since A, parallel transported to r via the path α-β, yields C. On the other hand, if we parallel transport A to r along path γ, the result is instead D, so one might equally say that A 'points in the same direction' as D. But of course this means trouble, for C and D, which live in the same tangent space, certainly do not point in the same direction as each other.

The solution is to abandon any *absolute* comparison of direction between arrows at different points: there is no fact about which arrow at r points in

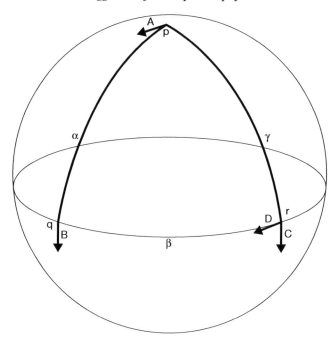

Figure 3. Parallel Transport on a Sphere

the same direction as A at *p*. But this is not to say that no comparison can be made, since one can compare A *as transported along a specified path to r* with arrows at *r*. We have been tricked into thinking that distant parallelism is an absolute matter because we live in a space that is very nearly flat, so for all practical purposes an arrow parallel transported from one place to another will yield the same result no matter which path is taken to connect them. In a perfectly flat Euclidean space, the result is exactly the same no matter which path is taken (Figure 4), but even in this case the *metaphysics* of parallelism is not that of a metaphysically pure relation: two distant arrows are only parallel in virtue of the affine connection along paths which connect their locations, even if the result of the parallel transport will be the same along any path.

Just as in the case of distance, the deep ontological structure of comparison between arrows at different points involves reference to continuous paths that connect the points. Arrows at locations that are not part of a single connected space can no more be compared than the locations themselves can have any distance between them.

Suggestions for Deep Metaphysics 93

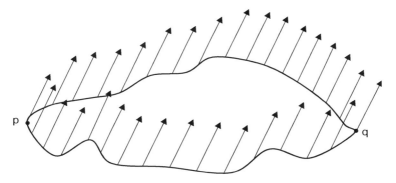

Figure 4. Parallel Transport in a Euclidean Space

In sum, the familiar notion of parallelism is not a metaphysically pure relation, either internal or external. Even if it is possible for point *p* with its tangent space and point *r* with its tangent space to be the only things that exist (with no additional space connecting them), still there would be no possible world which contains only them and in which any comparison of direction at all can be made between the elements in the two tangent spaces.

We saw above that the ontological structure of the world cannot be explicated if the only metaphysically real feature the universe has is its cardinality. Lewis insists that we need some notion of sparse properties if we are to make sense of the notion of *similarity of entities*: similarity will be a matter of sharing universals, or being composed from duplicate tropes, or being members of some primitively natural set. According to any of these three accounts of similarity, similarity itself is a metaphysically pure relation: two items can share a universal, or contain duplicate tropes, or be common members of a primitively natural set, even if they happen to be the only two items in existence. Similarity or dissimilarity of direction between arrows in a space does not fit any of these models. In one sense, one might say that there is no similarity relation between directions at different points at all. But in a practical sense, comparisons can be made between directions at different points so long as parallel transport along the paths that anyone is likely to use give results which are, within common margins of error, the same. So we can, for all practical purposes, understand the truth conditions for claims about parallelism even where there is, strictly speaking, no absolute, metaphysically pure relation of 'pointing in the same direction' at all.

We are now in a position to understand the bearing of gauge theories on deep metaphysics. Gauge theories apply exactly the sort of structures that we have used to explicate comparison of directions to other sorts of fundamental physical 'properties' or 'magnitudes'. Take, for example, chromodynamics, the theory of the force that binds quarks together. The easiest way to begin to describe the theory employs the language of universals: there are three color 'charges' ('red', 'blue', and 'green'), which are analogous to, e.g. positive and negative electric charge. There is a force produced between colored particles, like the electric force. This force is mediated by the gluons. And so on.

If one were to take this language at face value, one would likely conclude that the three colors are metaphysically pure properties. Whether explained by reference to the universal 'red', or to duplicate tropes of red, or by the natural set of red things, one would be inclined to accept that there is an absolute fact about whether any given quark is red or (because there may be some difficulty understanding how our terms can manage to refer determinately to one color rather than another) at least a fact about whether any two quarks have the same color charge or not. If the various charges were metaphysically pure properties, then one would be forced to conclude that the relations of sameness and difference of charge are metaphysically pure internal relations.

But chromodynamics is a gauge theory, which means that at base color charge is treated completely analogously to directions. The mathematical gadget one uses to formulate the theory is called a *fiber bundle*. In order to build a fiber bundle, one begins with a *base space*, which is going to be space-time. To every point of the base space, one then associates a *fiber* with a given geometrical structure. Figure 2 represents a sort of fiber bundle, with the surface of the sphere as the base space and the tangent space at each point as the fiber. Indeed, any manifold has an associated *tangent bundle* of just this form. In the case of the tangent bundle the geometrical structure of the fiber is fixed by the geometrical structure of the base space: the tangent space for a point in a two-dimensional space is itself a two-dimensional vector space. But in the general case, the structure of the base space puts no constraints on the structure of the fiber. We could, for example, choose a fiber of any dimensionality and any topology. We could associate a *circle* with each point in the base space (this is done in the theory of electromagnetism), or a torus, or a twelve-dimensional sphere. Since we are trying to model the three colors of chromodynamics, we will choose a sphere as the fiber. We will represent

Suggestions for Deep Metaphysics

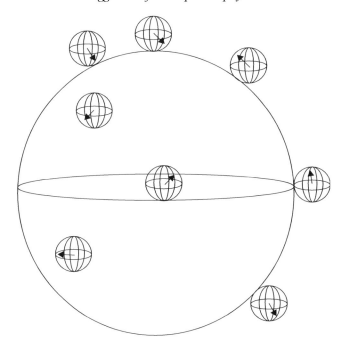

Figure 5. A Fiber Bundle on the Sphere

the 'color state' at a point by an arrow which points some direction in the sphere, as in Figure 5:

The idea that there are three distinct quark colors is reflected in the choice of the sphere as the fiber: there are three orthogonal directions that the arrow can point. (The full set of directions are available as color states since in addition to red, blue, and green one has all of the quantum superpositions of these states.[7]) So at this stage of the construction, each point in the base space has a space of possible color states associated with it, but we have no means of comparing the states at *different* points with each other. In order to do this, we need to add something more, something which, intuitively, ties together the fiber at every point with the fibers of points that are infinitesimally close. The something more we need is a *connection* on the fiber bundle. Once we have

---

[7] The actual mathematics of chromodynamics is being misrepresented here, since the actual gauge group for QCD is SU(3), not O(3), but these details are not important for the fundamental metaphysical morals.

a connection, we can do exactly (and only!) the sort of comparison we did with directions: given points p and q in the base space *and a continuous path connecting them*, we can 'parallel transport' a vector from the fiber over p along the path to the fiber over q and compare the transported vector from p to the vector at q. And just as for directions, the results of the comparison will in general depend upon the particular path chosen: there is no path-independent fact about whether vectors in different fibers are 'the same' or 'different'.[8]

So suppose we somehow take one of the fibers and arbitrarily choose three orthogonal directions to be the blue, red, and green directions (or alternatively, suppose we could pick out a quark and dub the color state it currently has as 'blue'). We could now characterize the actual state at that fiber in terms of the color vocabulary. But there would still not be any metaphysical fact about the color of any *other* quark at any other point, or whether any two quarks have the same or different colors. And most importantly, the physics does not *need* any such absolute comparison of quark colors: the physics can be stated entirely in terms of the connection on the fiber bundle. (Indeed, in gauge theory the electromagnetic field or the gluon field turns out just to *be* the connection on a fiber bundle.)

If we adopt the metaphysics of the fiber bundle to represent chromodynamics, then we must reject the notion that quark color is a universal, or that there are color tropes which can be duplicates, or that quarks are parts of 'natural sets' which include all and only the quarks of the same color, for there is no fact about whether any two quarks are the same color or different. Further, we must reject the notion that there is any metaphysically pure relation of comparison between quarks at different points, since the only comparisons available are necessarily dependent on the existence of a continuous path in space-time connecting the points. So it seems that there are no color properties and no metaphysically pure internal relations between quarks.[9] And if one believes that fundamental physics is the place to look for the truth about universals (or tropes or natural sets), then one may find that physics is telling us there are no such things.

---

[8] If the connection were (in the appropriate sense) flat, the result of transporting a vector might be the same no matter which path is chosen. But this sort of path *invariance* should not fool us into thinking that the comparison is metaphysically path *independent*: comparisons can only be made if there are paths connecting the points.

[9] One could suggest that there are still color properties, but that every point in the space-time has its own set of properties, which cannot be instantiated at any other point. But since such point-confined properties could not underwrite any notion of similarity or dissimilarity (just as abundant properties cannot), it is hard to see what would be gained by adopting the locution.

3. SOME IMPLICATIONS OF FIBER BUNDLES AS ONTOLOGICAL STRUCTURE[10]

The notion of a fiber bundle can be seen as a generalization of the mathematical structure one would use to represent universals. Suppose, for example, there were a generic universal, 'color', which had many particular specific instances, which we will call 'hues'. And suppose there were an unproblematic topology on the set of hues: hues could be nearer or further from one another, perhaps, in different respects (intensity, saturation, tint, etc.). Then the set of specific instances would form a sort of space, which we may call 'color space'. And every colored object, which would instantiate one specific hue, would be associated with a particular point in the color space. Specifying the color distribution of objects in the world would therefore amount to specifying a *function* from the base space (in this case points of space-time) to the color space (Figure 6). Two points in the base space that map to the same point in the color space would be exactly the same color. This would correspond to the two points instantiating the same universal, or containing identical tropes, or belonging to the same natural set.

There is evidently a way to reconceptualize this picture as an instance of a fiber bundle: make as many copies of the color space as there are points in the base space, and glue a copy on each. The identity of hues when comparing different fibers would be settled and absolute, and it would not matter what the topology of the base space was: it could even be disconnected. Note that in this picture there are necessarily many distinct *continuous functions* of color

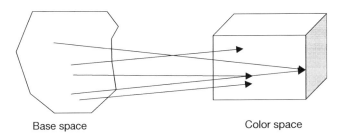

Figure 6. Mathematical Representation of a Universal

---

[10] This section follows the presentation in Chapter 15 of Penrose 2005, which is highly recommended as a more thorough introduction

on the base space. There are, for example, all of the *constant functions* that one can get by mapping each point in the base space to the same point in the color space, and there would be many other continuous functions beside.

To generalize this picture to the less constrictive notion of a general fiber bundle, one begins by erasing the particular identities of the hues in the fibers. Each fiber, though, will retain a sort of *geometrical structure*, as it were, a generic structure that the fibers all share. This structure *is* a universal, and so provides an exception to our overgeneralization that there are no metaphysically pure external relations: different fibers can have the same geometrical structure. An example of such a generic structure would be an n-dimensional vector space: the tangent spaces of Figure 2, for example, are all two-dimensional vector spaces. These vector space admit of *symmetries*, transformations that leave the generic structure unchanged. For example, the structure of the two-dimensional vector space is unchanged under rotations or uniform stretching. These symmetries will form a group, the symmetry group of the fiber. A vector space also will have a distinguished vector, the zero vector, that plays a unique role in the space.

As we have seen, the connection on the fiber bundle specifies the way that nearby fibers (i.e. fibers over nearby points in the base space) are 'glued together'. For this gluing to make mathematical sense, the *generic* structure of the fibers must match up: if the fiber is a vector space, for example, the zero vectors in different fibers must get glued onto each other. But since the fibers admit of symmetries, it follows that the fibers can be 'matched up' in different ways: applying a symmetry transformation to one will not change its generic structure. Unlike our original example of the color space, where the identities of the specific hues allow for only one way for different fibers to 'match up', we now can see how the connection is needed to specify something like an identification of elements in different fibers, but that identification is only relative to a path connecting the points in the base space: it is not absolute.

To see some consequences, let's consider one of the simplest non-trivial fiber bundles. Let the base space be a circle and the fiber be a one-dimensional vector space. The vector space has a zero vector, of course, and an important symmetry that maps each vector to its negative. One can think of the vector space as like the real line, but with no intrinsic labeling of one direction from zero as positive and the other as negative (nor is there an absolute scale). One way to form a fiber bundle is the obvious one: connect the fibers to one another in such a way as to form a *cylinder* (Figure 7). The analog to a continuous *function* on the base space is a continuous *section* of the bundle:

Suggestions for Deep Metaphysics

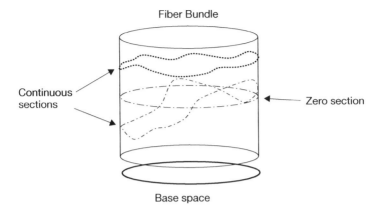

Figure 7. Fiber Bundle over a Circle

a specification of a particular element in each fiber that is continuous in the bundle. This particular bundle evidently admits of many different continuous sections, including the zero section that runs through all of the zero vectors. There is, as yet, nothing that counts as a *constant* section: recall that we have no general way to make sense of an element in one fiber being the same or different from an element in another (the zero vector being the exception). So the figure could be misleading: which sections appear as 'horizontal slices' is just a convention of the diagram. But there are evidently many continuous sections that nowhere intersect the zero section.

Due to the symmetry of the fiber, though, there is a topologically distinct way to form a bundle with this base space and fiber: by putting a 'twist' in the bundle, we can make a Möbius strip instead (Figure 8). There are still many continuous sections, including the zero section, but the difference in topology makes for a vast difference in the sorts of continuous sections that exist. In particular, it is easy to see that every continuous section must intersect the zero section in at least one place, a fact that does not obtain in the cylinder.

The situation is therefore much like that remarked above regarding the Triangle Inequality. Taking distance relations as ontological primitives demands the imposition of the Triangle Inequality as an additional constraint if the distance relations are all to be embedded in a single common space. Making distance a derivative notion, defined in terms of path length, makes the Triangle Inequality come out as a necessary truth without any further work. Similarly, if one's basic mathematical representation of a physical state of affairs is a continuous *function*, then there will necessarily be many

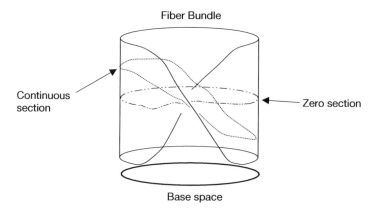

Figure 8. A Twisted Fiber Bundle

distinct, everywhere non-zero continuous functions available unless some further constraint is imposed. Using a continuous *section* of a fiber bundle may, depending on the structure of the bundle, yield a quite different set of possible physical states. There are even fiber bundles in which no continuous sections exist at all.[11]

In addition to the topological distinction between bundles illustrated by the cylinder and the Möbius strip, there are essential differences between bundles because of the *curvature* of the connection. In a bundle with a flat connection, parallel transport of a vector in a fiber around a closed curve (in the base space) will always bring the vector back completely unchanged. But if the connection is curved, this may not occur. Figure 3, the tangent bundle on a sphere, again illustrates the situation. Vector C parallel transported along the triangle $\beta-\alpha-\gamma$ returns to point r as vector D, rotated through a 90 degrees. The comparison between C and D is unproblematic since they are elements of the same fiber. This indicates that the curvature in the region enclosed by the triangle is non-zero: indeed it even allows us a way to define the average curvature in the region and hence, by a limiting process, the curvature at r.[12] But this curvature is not a Humean property or quality of the point r alone, it is rather a feature of the larger structure which is the fiber bundle.

---

[11] If the fiber is a vector space, then there will always be at least one continuous section: the zero section. But the fiber need not be a vector space. In the Clifford bundle, which has no continuous sections, the fiber is a circle. See Penrose 2005, pp. 334–8.

[12] Because the surface of the sphere is two-dimensional, the curvature at a point is just a scalar: in a higher-dimensional base space the curvature will be represented by a tensor.

Fiber bundles provide new mathematical structures for representing physical states, and hence a new way to understand physical ontology. For example, modern electromagnetic theory holds that what we call the 'electromagnetic field' just *is* the connection on a fiber bundle. Such an account evidently carries with it quite a lot of ontological commitments: there must be a base space, and internal degrees of freedom at each base point represented by a fiber, and a unified object that corresponds to all of the fibers with a connection. But if one asks whether, in this picture, the electromagnetic field is a *substance* or an instance of a *universal* or a *trope*, or some combination of these, none of the options seems very useful. If the electromagnetic field is a connection on a fiber bundle, then one understands what it is by studying fiber bundles directly, not by trying to translate modern mathematics into archaic philosophical terminology. If an electromagnetic field is a connection on a fiber bundle, then there are more things in heaven and earth than dreamt of in Plato's or Aristotle's philosophy. And surely this is to be expected: it would be a miracle if the fundamental ontological structure of the universe fit neatly in the categorical pigeonholes handed down to us from two millennia ago.

The fan of philosophical ontology (universals, tropes, natural sets, or whatever) can no doubt find a way to jerry-rig an object that can do the same mathematical work as a fiber bundle, just as the determined relationist can accommodate the Triangle Inequality by positing it as a separate law of nature (or of ontology?). The question is what the point of such a exercise could possibly be. Fiber bundles have their own, interesting structure, a structure that does not correspond to the traditional philosophical vocabulary. And the structure of the physical world might be that of a fiber bundle. So instead of a rearguard operation in defense of philosophical tradition, philosophers would do best to try to understand the structure on their own terms. If they do not translate well into the categories in which philosophical debates have taken place, so much the worse for the philosophical debates.

4. LOOSE ENDS

We began with the notion that the grammar of natural language could be the source of the metaphysical picture of substance and universal. But the theory of universals does more than mimic the structure of predication, it also provides the foundation of an account of similarity and difference between objects. Since it is hard to see how to give an adequate account of the world without some account of similarity and difference, we have been examining a way to provide a sort of comparison without admitting a metaphysically pure internal relation of comparison. But have we really rejected the metaphysics of substance and universal entirely?

We have not even attempted to eliminate any notion of substance. Given the dependence of so much of the fiber-bundle structure on objects inhabiting a single connected space-time, one might be inclined to go Spinozistic: there is but one substance, and it is the whole of connected space-time. We naively think that parts of the space-time have intrinsic properties *that are intrinsically comparable to each other*, and so we naively think that parts of space-time (or objects in different parts of space-time) are themselves metaphysically independent substances, but in this we are mistaken. Of course, going Spinozistic is really a Pyrrhic victory for a metaphysics of substance and universal: if there is only *one* substance and every physical fact corresponds to a distinct property of *it*, then we are getting no structural insight at all into the nature of the world by being informed of which substances exist. One can retain the vocabulary of substance and universal, but the cash value is nearly nil.

Furthermore, we have not really eliminated all notions of universals or metaphysically pure internal relations. When we constructed the fiber bundles, we began by associating to each point in the base space fibers *with the same geometrical structure.* Fiber bundle theory may imply that there is no absolute comparison between physical states like color charge at different points, but there are absolute identities of form at a more abstract level. Or to put the point in a somewhat tendentious way, we might say that although there are no universals that correspond to *matter* or to *physical magnitudes*, there are geometrical universals of pure form. But since metaphysicians like Armstrong have focused on examples like electric charge and mass in explicating the theory of universals, eliminating these requires a wholesale revision of that picture of universals.

We also have not ruled out the existence of *functionally defined properties*, i.e. properties that are specified by quantifying over possible physical manifestations, such as computational structure. For example, even if two objects in disconnected space-times cannot have comparable physical characteristics, it appears possible that they could both be instantiations of the same Turing machine, since the specification of a Turing machine is not given in terms of first-order physical vocabulary at all. However, we recognize that such functionally defined properties are ontological parasites: an entity can only instantiate them in virtue of having some non-functionally defined structure. Our concern has been the metaphysical status of this ground-level structure, which is commonly identified as physical structure.

Finally, we should note that adopting the metaphysics of fiber bundles invalidates a set of modal intuitions that have been wielded by David Lewis under the rubric of the Principle of Recombination. According to Lewis, Hume taught us that the existence of any item puts no metaphysical constraints on what can exist adjacent to it in space. This invites a cut-and-paste approach to generating metaphysical possibilities: any object could in principle be *duplicated* elsewhere, immediately adjacent to the duplicate of any other item (or another duplicate of itself). The notion of duplication presupposes universals (or tropes or natural sets), and for Lewis the relation of duplication can hold even between regions in disconnected space-times (i.e. in different possible worlds). Duplication is supposed to be a metaphysically pure internal relation between items. But from the point of view of fiber bundle theory, it makes no sense to 'copy' the state of one region of space-time elsewhere even in the same space-time, much less in a disjoint space-time. There is no metaphysical copying relation such as the Principle of Recombination presupposes. So modal intuitions based on the Principle of Recombination ought to be suspect.

Contemporary physics provides us with a fundamental ontological picture that is distinct from the philosophical views currently on offer. As speculative philosophers, we should be interested in this alternative if only because it forces us to stretch our conceptual boundaries. But more than that: given the way that the fiber bundle theories came to be developed, we have at least some reason to think that they, and not the theory of universals or its kindred, are true.

4
On the Passing of Time

Metaphysics is ontology. Ontology is the most generic study of what exists. Evidence for what exists, at least in the physical world, is provided solely by empirical research. Hence the proper object of most metaphysics is the careful analysis of our best scientific theories (and especially of fundamental physical theories) with the goal of determining what they imply about the constitution of the physical world.

The foregoing theses strike me as incontestable. If one accepts them, the project of metaphysics takes on a form rather different than that commonly practiced today. Long gone, of course, are logical empiricist attempts to reduce all meaningful assertions to claims about sense experience. The metaphysical irreducibles are to be provided by physics—quarks, electrons, and space-time, for example—rather than by 'epistemological priority'. The horror inspired by 'unverifiable' propositions completely dissipates, an attitude nicely attested by the physicist J. S. Bell when discussing interpretations of the quantum theory:

> The 'microscopic' aspect of the complementary variables is indeed hidden from us. But to admit things not visible to the gross creatures that we are is, in my opinion, to show a decent humility, and not just a lamentable addition to metaphysics. (Bell 1987, pp. 201–2)

Empiricists subordinated metaphysics to epistemology via the empiricist theory of meaning. For Hume, the simple elements of any idea had to be provided by experience, so talk of the in-principle unobservable must be empty of cognitive content. Similarly for the logical empiricists, albeit with the emphasis shifted first from ideas to sentences and then, with Hempel and Quine, to complete theoretical systems. Since the collapse of the empiricist

---

Thanks to David Albert, Frank Arntzenius, John Earman, Adam Elga, Shelly Goldstein, Ned Hall, Barry Loewer, Huw Price, and Brad Weslake for, in some cases, vigorous opposition, and in others, sympathetic support. You know who is who. The first section of this paper appeared as Maudlin 2001.

criteria of meaning, philosophers have no longer been trying to reduce the theoretical vocabulary of physics to middle-sized-thing-language, and there is no good reason to try to reinvigorate that moribund program.

But the ghost of the empiricist program still haunts metaphysics. It manifests itself most obviously in the thesis of Humean Supervenience, which animates the work of David Lewis. Lewis claims that his Humeanism springs exactly from a desire to protect physics from incursions by philosophers:

> Is this [namely Humean Supervenience] materialism? – no and yes. I take it that materialism is metaphysics built to endorse the truth and descriptive completeness of physics more or less as we know it; and it just might be that Humean supervenience is true, but our best physics is dead wrong in its inventories of the qualities. Maybe, but I doubt it. Most likely, if Humean supervenience is true at all, it is true in more or less the way that present physics would suggest …
> Really, what I uphold is not so much the truth of Humean supervenience but the *tenability* of it. If physics itself were to teach me that it is false, I wouldn't grieve. (Lewis 1986a, p. xi)

But the practical results of embracing Humean Supervenience are not the same as turning to scientific practice for the raw materials from which clear ontology is refined. For example, Lewis's Humeanism impels him to search for a reductive analysis of laws of nature (the so-called Mill–Ramsey–Lewis best systems theory). But nothing in *scientific practice* suggests that one ought to try to reduce fundamental laws to anything else. Physicists simply postulate fundamental laws, then try to figure out how to test their theories; they nowhere even attempt to analyze those laws in terms of patterns of instantiation of physical quantities. The practice of science, I suggest, takes fundamental laws of nature as further unanalyzable primitives. As philosophers, I think we can do no better than to follow this lead.

Accepting unanalyzable primitives into one's ontology may seem like philosophical dereliction of duty: after all, much of metaphysics is devoted to providing analyses or reductions. But the exact point of these reductions is not always clear: ham-fisted appeals to Ockham's Razor or a hankering for Quinean desert landscapes do not stand up as justifications for accepting the most bare-bones ontology that can be contrived. To take a single example: one can try to eliminate fields from one's ontology by concocting an action-at-a-distance theory, but even if the concoction can be achieved, why think that the resulting theory is more likely to be true than the field theory one began with? Less is not always more, and certainly less is not always more justified. On the other hand, if every entity is to be allowed in as an irreducible primitive, the metaphysician's work is quickly, and uninspiringly, done.[1]

A general, abstract account of when ontological reductions should be pursued and when a bare posit should be made would no doubt be a Good Thing. I am unsure whether any such general account is possible, and am certainly unable to provide one here. In its stead, I can only offer a profession of what Nelson Goodman would call my 'philosophic conscience' (Goodman 1983, p. 32). I believe there is a fundamental physical state of the world. One may roughly think of that state as consisting of electrons and quarks and photons and so on distributed throughout space-time, but the work of understanding just what electrons and quarks and space-time are remains undone. (Part of that state appears to include a quantum wavefunction, which is not like electrons, or fields, or space-time, and deserves its own careful scrutiny.[2]) I am a substantivalist about space-time insofar as I do not think that spatio-temporal facts can be reduced to relational facts about material bodies as more or less classically conceived. Since it likely will turn out that in our best theories neither electrons and quarks nor space-time can be more or less classically conceived, many current philosophical disputes about space and time are likely to need revisiting in the future.

In addition to the physical state of the universe, I believe in fundamental physical laws.[3] Facts about what the laws are cannot be reduced to, or analyzed in terms of, other sorts of facts. My philosophical conscience dictates that ultimately the physical state and the fundamental physical laws are all there are in the inanimate realm: all astronomical or chemical or meteorological facts supervene on these. Insofar as counterfactual and causal claims have determinate truth conditions, the ontology that underwrites the truth values of these claims is just the physical state and the fundamental physical laws.[4] There are no ontologically additional biological or chemical or economic laws, there are no further brute facts about causation or counterfactuals or dispositions.

---

[1] At least at one level. Inflating ontology can, in some contexts, pose severe questions of *choreography* among the entities. If one thought that, say, typewriters were not composite entities among whose parts are keys, it would be puzzling why one cannot manage to use a typewriter without also touching some keys.

[2] The wavefunction appears to be a section of a fiber bundle: see 'Suggestions from Physics for Deep Metaphysics', Chapter 3, this volume.

[3] See 'A Modest Proposal Concerning Laws, Counterfactuals and Explanation' Chapter 1, this volume.

[4] See 'Causation, Counterfactuals, and the Third Factor', Chapter 5, this volume.

The qualifier 'inanimate' appears above as a bow to the mind-body problem. I admit that the evident existence of subjective mental states is neither obviously part of, nor reducible to, physical state and physical law. But I do not think that all ontological analyses need be held hostage to this conundrum. In particular, investigation of the physical ontology can proceed so long as the physical world contains plausible *de facto* correlates of subjective mental states, such as the notorious firing of C-fibers for pain. Identifying *de facto* correlates leaves a lot of work to be done, such as identifying the generic features of these correlates in virtue of which they are correlated to mental states, explaining the connection between the third-person and first-person description, and so on. In this sense, the Newtonian (or Democritean) world-picture does not solve the mind-body problem but still can proceed without such a solution. In actual practice, the Newtonian need only derive from the theoretical apparatus states that correspond to what we take to be the manifest observable structure of the world: a Newtonian derivation of a parabolic trajectory for a thrown rock can be tested in obvious ways in the lab with no notice of conscious states at all. But a more fastidious Newtonian could, for example, trace a set of interactions between the rock and the firing of neurons in the brains of the experimenters that would suffice to underwrite the reliability of the laboratory investigation.

The general ontological picture just adumbrated is by no means popular. The notion that laws of nature are irreducible, for example, has few adherents. Furthermore, when considering the exact nature of the physical state of the universe, I find that I am strongly inclined to a view that strikes many of my colleagues as lunacy. I believe that it is a fundamental, irreducible fact about the spatio-temporal structure of the world that time passes.

Skepticism about the view that time passes (in the sense of that claim that I intend to defend) takes many forms. At its most radical, the skepticism appears as blank bewilderment about what I could possibly mean to say that time passes. Since I take the passage of time to be a fundamental, irreducible fact, there is some difficulty how to respond to this bewilderment. I cannot explain what I mean by paraphrasing or analyzing the notion of time's passage in terms that do not already presuppose the notion. But the situation is not altogether desperate. I can indicate at least some features of the passage of time that serve to distinguish time from space, and that serve to distinguish this problem from other problems. Usually this is sufficient to convince my interlocutors that whatever it is I believe in, it is something that they do not.

The passage of time is an intrinsic asymmetry in the temporal structure of the world, an asymmetry that has no spatial counterpart. It is the asymmetry that grounds the distinction between sequences which run from past to future and sequences which run from future to past. Consider, for example, the sequence of events that makes up an asteroid traveling from the vicinity of Mars to the vicinity of the Earth, as opposed to the sequence that makes up an asteroid moving from the vicinity of Earth to that of Mars. These sequences might be 'matched', in the sense that to every event in the one there corresponds an event in the other which has the same bodies in the same spatial arrangement. The topological structure of the matched states would also be matched: if state B is between states A and C in one sequence, then the corresponding state B* would be between A* and C* in the other. Still, going from Mars to Earth is not the same as going from Earth to Mars. The difference, if you will, is how these sequences of states are oriented with respect to the passage of time. If the asteroid gets closer to Earth as time passes, then the asteroid is going in one direction, if it gets further it is going in the other direction. So the passage of time provides an innate asymmetry to temporal structure.

In 'An Attempt to Add a Little Direction to "The Problem of the Direction of Time" ', John Earman describes a view he calls *The Time Direction Heresy* as follows:

> It states first of all that if it exists, a temporal orientation is an intrinsic feature of space-time which does not need to be and cannot be reduced to nontemporal features, and secondly that the existence of a temporal orientation does not hinge as crucially on irreversibility as the reductionist would have us believe. (Earman 1974, p. 20)

Earman himself does not unequivocally endorse the Heresy, but does argue that no convincing arguments against it could be found, at that time, in the very extensive literature on the direction of time. Over three decades later, I think that this is still the case, and I want to positively promote the Heresy. The arguments against it that I will consider are largely disjoint from those surveyed by Earman, so this Chapter can be seen as a somewhat more aggressive companion piece to his. The intrinsic, irreducible temporal orientation of the Heresy corresponds to a specification of the direction in which time passes.

The belief that time passes, in this sense, has no bearing on the question of the 'reality' of the past or of the future. I believe that the past is real: there are facts about what happened in the past that are independent of the present state of the world and independent of all knowledge or beliefs about the past. I similarly believe that there is (i.e. will be) a single unique future. I know what it would be to believe that the past is unreal (i.e. nothing ever happened, everything was just created *ex nihilo*) and to believe that the future is unreal (i.e. all will end, I will not exist tomorrow, I have no future). I do not believe these things, and would act very differently if I did. Insofar as belief in the reality of the past and the future constitutes a belief in a 'block universe', I believe in a block universe. But I also believe that time passes, and see no contradiction or tension between these views.

The inexact use of the phrase 'block universe theory' can systematically distort and confuse discussions of the passage of time. Earman, for example, would cede pride of place to no one in his commitment to a single, four-dimensional relativistic space-time, but for all that he rightly considers that the Heresy could be correct. Here is Huw Price on the terminology:

> Often this is called the *block universe view*, the point being that it regards reality as a single entity of which time is an ingredient, rather than as a changeable reality set *in* time. The block metaphor sometimes leads to confusion, however. In an attempt to highlight the contrast with the dynamic character of the 'moving present' view of time, people sometimes say that the block universe is *static*. This is rather misleading, however, as it suggests that there is a time frame in which the four-dimensional block universe stays the same. There isn't of course. Time is supposed to be included in the block, so it is just as wrong to call it static as it is to call it dynamic or changeable. It isn't *any* of these things, because it isn't the right sort of entity—it isn't an entity *in* time, in other words.
>
> Defenders of the block universe deny that there is an objective present, and usually also deny that there is any objective flow of time. (Price 1996, pp. 12–13)

I am one of those unusual defenders of the block universe who does not deny that there is any objective flow of time. The four-dimensional universe is a single entity of which the *passage* of time, in one particular direction, is an ingredient. We will return to Price's objections to this doctrine presently.

The passage of time is deeply connected to the problem of the direction of time, or time's arrow. If all one means by a 'direction of time' is an irreducible intrinsic asymmetry in the temporal structure of the universe, then the passage of time implies a direction of time. But the passage of time connotes more than just an intrinsic asymmetry: not just any asymmetry would produce passing. Space, for example, could contain some sort of intrinsic asymmetry, but that alone would not justify the claim that there is a 'passage of space' or that space passes. The passage of time underwrites claims about one state 'coming out of' or 'being produced from' another, while a generic spatial (or indeed a generic temporal) asymmetry would not underwrite such locutions.

Not infrequently, the notion of the passage of time is discussed under the rubric 'time's flow', or 'the moving now'. These locutions tend to be favored by authors intent on dismissing the notion, and certainly subserve that purpose. For time to literally 'move' or 'flow', they say, there must be some second-order time by means of which this movement or flow is defined, and off we go up an infinite hierarchy of times. Quite so. Except in a metaphorical sense, time does not move or flow. Rivers flow and locomotives move. But rivers only flow and locomotives only move because time passes. The flow of the Mississippi and the motion of a train consist in more than just the collections of instantaneous states that have different relative positions of the waters of the Mississippi to the banks, or different relative positions of the train to the tracks it runs on. The Mississippi flows from north to south, and the locomotive goes from, say, New York to Chicago. The direction of the flow or motion is dependent on the direction of the passage of time. Common locutions speak of time passing and things in time moving or flowing or changing. Given the essential role of the passage of time in understanding the notion of flow or motion or change, it is easy to see why one might be tempted to the metaphor that time itself flows. (Interestingly, there are few examples of extending the notion in the other way: time is the thing that passes. No doubt this is because the passage of time is not explicated by means of any other more primitive notion.) But in order to avoid such confusions, I will stick to saying only that time passes.

My primary aim in this Chapter is to clear the ground of objections to the notion that time, 'of itself, and from its own nature, flows equably without relation to anything external'. The phrase, of course, is Newton's characterization of 'absolute, true, and mathematical time' in the Scholium to Definition 8 of the *Principia*, and, properly understood and updated to fit Relativity, I fully endorse it. To forestall some bad objections, I would, myself, replace 'flows equably' with 'passes', but the basic claim is right on target.

There are three sorts of objections to the passage of time, which we may group as logical, scientific, and epistemological. Logical objections contend that there is something incoherent about the idea of the passage of time *per se*: conceptual analysis can show the idea to be untenable or problematic. Scientific objections claim that the notion of the passage of time is incompatible with current scientific theory, and so would demand a radical revision of the account of temporal structure provided by physics itself. Epistemological objections contend that even if there were such a thing as the passage of time, we could not know that there was, or in which direction time passes. The second and third groups are not so distinct, and my discussion may run these sorts of arguments together.

For expository purposes, we are fortunate to have a text that touches on all of these sorts of objections. The passage occurs in Huw Price's book *Time's Arrow and Archimedes' Point*, under the heading 'The Stock Philosophical Objections about Time' (Price 1996, pp. 12–16). The heading is apt. These are, indeed, the very objections that arise spontaneously in any discussion of these matters, and, with one exception, they are not original to Price himself. He has, however, done us a service in collecting and presenting them in such a clear way, and I will use his exposition as a guide.

I should also clearly announce my disagreement with Price. It is not that Price thinks these arguments decisively refute the idea of the passage of time while I think that the case is not closed. Price does not consider these arguments decisive, but only relatively strong and plausible:

In making these assumptions [namely that there is a block universe *and* that there is no 'objective flow of time'] I don't mean to imply that I take the arguments for the block universe view sketched above to be conclusive. I do think that it is a very powerful case, by philosophical standards. (Price 1996, p. 15)

In contrast, I think that the arguments rehearsed have no force whatsoever against the existence of an 'objective flow of time', i.e. the view that time, of itself and by its own nature, passes.

1. LOGICAL ARGUMENTS

The stock logical objection is presented by Price as follows:

if it made sense to say that time flows then it would make sense to ask how fast it flows, which doesn't seem to be a sensible question. Some people reply that time flows at one second per second, but even if we could live with the lack of other possibilities, this answer misses the more basic aspect of the objection. A rate of seconds per second is not a rate at all in physical terms. It is a dimensionless quantity, rather than a rate of any sort. (We might just as well say that the ratio of the circumference of a circle to its diameter flows at π seconds per second!) (ibid. 13)

I think the 'more basic' aspect of this problem is indeed an original contribution of Price. It is, in any case, new to me. Let's deal with the original objection first.

Let's begin by considering the logic of rates of change. If something, e.g. a river, flows, then we can indeed ask how fast it flows (how many miles per hour, relative to the banks). To ask how fast a river flows is to ask how far the water in it will have gone when a certain period of time has passed.[5] If the Mississippi flows at 5 mph to the south (and maintains a constant rate), then after an hour each drop of water will be 5 miles further south. It will be 5 miles further from Canada and 5 miles closer to the equator.

On this basis, if we ask how fast time flows, i.e. how fast time passes, we must mean to ask how the temporal state of things will have changed after a certain period of time has passed. In one hour's time, for example, how will my temporal position have changed? Clearly, I will be one hour further into the future, one hour closer to my death, and one hour further from my birth. So time does indeed pass at the rate of one hour per hour, or one second per second, or 3,600 seconds per hour...

What exactly is supposed to be objectionable about this answer? Price says we must 'live with the lack of other possibilities', which indeed we must: it is necessary and, I suppose, a priori that if time passes at all, it passes at one second per second. But that hardly makes the answer either unintelligible or meaningless. Consider the notion of a fair rate of exchange between currencies. If one selects a standard set of items to be purchased, and has the costs of the items in various currencies, then one may define a fair rate of exchange between the currencies by equality of purchasing power: a fair exchange of euros for dollars is however many euros will purchase exactly what the given amount of dollars will purchase, and similarly for yen and yuan and so on. What, then, is a fair rate of exchange of dollars for dollars? Obviously, and necessarily, and a priori, one dollar per dollar. If you think that this answer is meaningless, imagine your reaction to an offer of exchange at any other rate. We do not need to 'live with' the lack of other possibilities: no objectionable concession is required.[6]

---

[5] One of course begins by defining rates of flow in terms of amount of change per unit time, assuming the flow to be constant. For objects whose rates change, like water in a river, the instantaneous flow is the limit as the relevant time approaches zero.

[6] For some reason I cannot quite comprehend, this rather trivial observation has preoccupied several of those who have read this essay. So to be absolutely clear: this is a comment about the concept *fair rate of exchange*. An exchange of dollars for euros such that the amount of dollars does

What of Price's 'more basic' objection? This, I fear, is just a confusion. A rate of one second per second is no more a dimensionless number than an exchange rate of one dollar per dollar is free of a specified currency. Price seems to suggest that the units in a rate can 'cancel out', like reducing a fraction to simplest terms. Any rate demands that one specify the quantities put in the ratio: without the same quantities, one no longer has the same ratio.

Suppose, for example, we are bartering floor tiles for licorice sticks. The agreed rate of exchange might be given in square feet (of tile) for feet (of licorice).[7] We are exchanging square feet of one thing for feet of another. It is simply a mistake to think we can 'cancel out' one of the feet in each quantity and say that the exchange is really in units of feet rather than square feet per foot. Of course, the rate will *transform* like a linear measure if we decide to change units: expressing our barter in terms of square inches of tile for inches of licorice, we multiply square feet by 144 and feet by 12. The real number used to express the ratio (i.e. the real number which stands in the same ratio to unity as the amount of tile stands to the amount of licorice, in the given units) will be multiplied by 12, just as if we were simply transforming a linear measure from feet to inches. But still the units of the barter are square feet per foot, not feet.

Similarly, π is defined as a ratio of a length (of the circumference of a Euclidean circle) to a length (of the diameter). The ratio is length to length: length does not 'cancel out'. There is, of course, also a real number (similarly called π, but don't get confused) that stands in the same ratio to unity as the circumference of a Euclidean circle stands to its diameter. That real number is dimensionless, but it plays no role in the definition of π.

π itself is defined independently of any *unit* of length. If one introduces a unit of length, then one can form a fraction whose numerator is the number of units in the circumference of a circle and whose denominator is the number of units in the diameter. This fraction (equal to the real number π) transforms like a dimensionless number when one changes units: it remains the same, so long as the same units are used for both measures. But still, length is involved in its definition, rather than weight or time or force. And the rate of passage

---

not have the same purchasing power as the amount of euros is not fair. Similarly, an exchange of some amount of dollars for another amount of dollars that does not have the same purchasing power is not fair. It follows analytically that the fair exchange rate of dollars for dollars is one for one.

[7] The exchange rate might also be given without any standard units at all, as *this much* tile (holding up a tile) to *that much* licorice (holding up a strand).

of time at one second per second is still a rate: it, unlike π, is a measure of how much something changes *per unit time*.

Price also mentions a logical objection to the notion of the *direction* of the passage of time:

> If time flowed, then—as with any flow—it would only make sense to assign that flow a *direction* with respect to a choice as to what is to count as a positive direction of time. In saying that the sun moves from east to west or that the hands of a clock move clockwise, we take for granted that the positive time axis lies toward what we call the future. But in the absence of some objective grounding for this convention, there isn't an objective fact as to which way the sun or the hands of the clock are 'really' moving. Of course, proponents of the view that there is an objective flow of time might see it as an advantage of their view that it does provide such an objective basis for the usual choice of temporal coordinate. The problem is that until we have such an objective basis we don't have an objective sense in which time is flowing one way rather than another. (ibid. 13)

This objection demands some untangling.

First, it will help greatly here to insist that properly speaking time passes, rather than flows, and that properly speaking anything that does flow only flows because time passes. The point about directionality of flow is then exactly correct: flows only have a direction because the asymmetry inherent in the passage of time provides temporal direction: from past to future. The natural thing is now to turn Price's Modus Tollens into a Modus Ponens: since there obviously is a fact about how the Mississippi flows (north to south) or how the hands of standard clocks turn (clockwise) there is equally a real distinction between the future direction in time and the past direction. The remark about choosing a convention for the 'positive direction of time' is a red herring: it is, of course, merely a convention that our clocks typically count up (i.e. indicate larger numbers as time passes) rather than count down. Nothing in the nature of the passage of time provides an 'objective basis' for that choice. A society that happens to build clocks that count down rather than up is not making any sort of mistake: attaching numbers to moments of time clearly requires purely arbitrary conventions. One who believes in the objective passage of time does not think there is an objective fact about which sort of clock is counting 'right' and which 'wrong', merely that there is an objective fact about which is counting up and which down. Up-counting clocks show higher numbers in the future direction, down-counting clocks lower numbers. To deny that there is an objective distinction between such clocks is to deny that there *is* any objective distinction between the future direction and the past, and that is precisely to beg the question.

This exhausts our examination of the logical objections to the passage of time.

2. SCIENTIFIC OBJECTIONS

Scientific objections to the passage of time stem from two sources. One is the spatio-temporal structure postulated by the Special and General Theories of Relativity, and the other is the so-called Time Reversal Invariance of the fundamental laws of physics. Let's take these in turn.

Price does bring up the theory of Relativity in this section, but not in the context of discussing the passage of time. He rather brings it up in defense of the block universe view: the view that past, present, and future are all equally real. The opponent of this view wants to give some special ontological status to the present, as opposed to the past or future, and this is hard to reconcile with a Relativistic account of space-time. Since I believe that the past, present, and future are all equally real I have no quarrel with Price here. But there are others who have claimed that the passage of time requires some spatio-temporal structure forbidden by Relativity.

Kurt Gödel, after a description of the familiar 'relativity of simultaneity' in the Special Theory, writes:

Following up the consequences of this strange state of affairs, one is led to conclusions about the nature of time which are very far reaching indeed. In short, it seems that one obtains an unequivocal proof for the view of those philosophers who, like Parmenides, Kant, and the modern idealists, deny the objectivity of change and consider change as an illusion or an appearance due to our special mode of perception. The argument runs as follows: Change becomes possible only through the lapse of time. The existence of an objective lapse of time, however, means (or at least is equivalent to the fact) that reality consists of an infinity of layers of 'now' which come into existence successively. But, if simultaneity is something relative in the sense just explained, reality cannot be split up into such layers in an objectively determined way. Each observer has his own set of 'nows', and none of these various systems of layers can claim the prerogative of representing the objective lapse of time. (Gödel 1949, pp. 557–8)

Gödel then goes on to describe his famous solution to the Einstein field equations, which not only cannot be 'split up into layers' (i.e. foliated into spacelike hypersurfaces) in a single objective way, it cannot be foliated into spacelike hypersurfaces at all (cf. Hawking and Ellis 1973, p. 170).

So we are left with two questions. First, does the passage of time imply a foliation of space-time, i.e. does it imply that the four-dimensional space-time structure is split into a stack of three-dimensional slices in an observer-independent way? Second, if it does, does this set the notion of passage of time in direct opposition to the account of space-time offered by our best scientific theories?

To the first question, I can find no justification for Gödel's blank assertion that the 'objective lapse of time' is 'equivalent' to the fact that reality is a stack of 'nows'. The passage of time provides, in the first instance, a fundamental objective distinction between two temporal *directions* in time: the direction from any event towards its future and the direction from any event towards its past. If we want to distinguish, for example, an asteroid going from Earth to Mars from an asteroid going from Mars to Earth, what do we need? We may focus completely on the world-line of the asteroid in question. Everyone agrees that one end of the world-line has Earth in the near vicinity of the asteroid, and the other end has Mars in the near vicinity: these facts do not require a foliation of the space-time. Does adding a foliation help *to any degree at all* in determining whether we have an Earth-to-Mars or a Mars-to-Earth trip? No. For even if we were to add the foliation, the crucial question of which events come first and which later would be unsettled. So the 'lapse of time' cannot be equivalent to the existence of a foliation.

Perhaps the 'lapse of time' is a foliation *plus* a specification of what the past-to-future direction is? This will certainly allow us to distinguish two trips from each other—but it is the direction, not the foliation, that is doing all the work. Give me the past-to-future direction and I'll tell you where the asteroid is going without reference to any foliation at all.

In a fully relativistic space-time, the obvious mathematical gadget one needs to distinguish one direction in time from another is not a foliation but an *orientation*. All relativistic models already employ orientable space-times: space-times in which the light-cones are divided into two classes, such that any continuous timelike vector field contains only vectors that lie in members of one of the classes. In order to account for the direction of flows or other motions, all we need to do is to identify one of these classes as the *future* light-cones and the other as the *past* light-cones. Once I know which set is which, I can easily distinguish a Mars-to-Earth asteroid from an Earth-to-Mars one. (See Earman 1974, section 2 for a more exact mathematical discussion.)

Having said that a foliation is not required for there to be a lapse of time, one should note that a foliation may well be forced upon us for purely physical reasons. In particular, although neither Special nor General Relativity posits a foliation of space-time, the only coherent, precise formulations of quantum theory we have seem to demand it, and the observed violations of Bell's inequality are very difficult to account for without the resources of such a foliation.[8]

What then of an orientation? Is that something which physics alone can do without, which is being added to our account of space-time for merely 'philosophical' reasons? Is it, further, something that would sit uneasily with the spatio-temporal structure posited by physics alone?

The treatment of this question is one of the most peculiar in the philosophical literature. The usual approach sets the problem as follows: the fundamental physical laws have a feature called 'Time Reversal Invariance'. If the laws are time reversal invariant, then it is supposed to follow that physics itself recognizes no directionality of time: it does not distinguish, at the level of fundamental law, the future direction from the past direction, or future light-cones from past light-cones. Therefore, it is said, any such distinction must be grounded not in fundamental law, or in the intrinsic nature of the space-time itself, but in contingent facts about how matter is distributed through space-time. The direction of time, we are told, is nothing but (e.g.) the direction in which entropy increases. The philosophical puzzle is then how to relate various other sorts of temporal asymmetry (the asymmetry of knowledge, or of causation, or of control) to the entropic asymmetry. Paul Horwich's *Asymmetries in Time* (1987) provides an example of this form.

This problematic is peculiar because it fails at every step. To begin with, the laws of physics as we have them (even apart from worrying about a coherent understanding of quantum mechanics) are not Time Reversal Invariant. The discovery that physical processes are not, in *any* sense, indifferent to the direction of time is important and well known: it is the discovery of the violation of so-called CP invariance, as observed in the decay of the neutral K meson. These decays are *not* invariant if one changes a right-handed for a left-handed spatial orientation (parity) and changes positive for negative charge (charge conjugation). According to the CPT theorem, any plausible quantum theory will be invariant under parity-plus-charge-conjugation-plus-time-reversal, so the violation of CP implies a violation of T. In short, the fundamental laws of physics, as we have them, *do* require

---

[8] See my 1994 and 1996 for details.

a temporal orientation on the space-time manifold. So the argument given above collapses at the first step.

How do philosophers respond to this difficulty? Horwich, having noted the problem, writes:

> However, this argument is far from airtight. First, the prediction has not been directly confirmed. And, even if it were true, it could turn out to be a merely *de facto* asymmetry, which does not involve time-asymmetrical laws of nature. Moreover neither the experimental nor the theoretical assumptions involved in the prediction are beyond question. For the frequency difference between the two forms of neutral K meson decay is not substantial and will perhaps be explained away. Anyway, the assumption that these processes are spatial mirror images may turn out to be false ... Finally, the so-called 'CPT theorem', though plausible, may be false. Since there are so many individually dubious assumptions in the argument, we may regard their conjunction as quite implausible. (Horwich 1987, p. 56)

There is a certain air of desperation about this passage. There is no dispute in the physics community about the reality or implications of this effect: Nobel prizes have been awarded both for the theoretical work and for the experimental verification of the effect. Insofar as philosophers of physics are looking for actual scientific results on, which to base ontological conclusions this is a clear case of the science testifying in favor of a temporal orientation. The only plausible reason for Horwich suddenly to turn skeptical is that the failure of T invariance spoils his argument.

There is a somewhat better response available. That would be to admit that the laws of physics are not Time Reversal Invariant, and that there is, indeed, a physical orientation to space-time, but to insist that the physical processes sensitive to this orientation (like neutral kaon decay) are too infrequent and exotic to explain the widespread and manifest distinction between the past direction and the future direction that we began with. We will examine the plausibility of this response in due course. But we should at least acknowledge that the admission of an orientation to space-time is not, *per se*, wildly at odds with present physical theory since present physical theory already admits one.

But let's set aside the observed violations of CP invariance. Even apart from these, it is not at all clear that the accepted laws of physics are Time Reversal Invariant *in a way that suggests that there is no intrinsic direction of the passage of time*. This point has been well argued by David Albert in his *Time and Chance* (2000, chapter 1). Here's the problem. The way that time reversal has been understood since the advent of electromagnetic theory is not: for every physically allowed sequence of instantaneous states, the same set of states in the reverse time order is also physically allowed. This can plausibly be argued to hold for Newtonian mechanics,[9] but beyond that one needs to do more than simply reverse the time order of the states: one has to perform a 'time reversal' operation on the instantaneous states themselves. So the theorem is now: for every physically allowable sequence of states, the inverse sequence of *time-reversed* states is also physically allowable. More precisely: if states $T_0, T_1, T_2, \ldots T_N$ are physically allowable *as time runs from T_0 to T_N in some direction*, then the sequence $T_N{}^*, \ldots T_2{}^*, T_1{}^*, T_0{}^*$ is also allowed, where the * represents the time reversal operation as applied to the states, and where time runs from $T_N{}^*$ to $T_0{}^*$ in the same direction as it runs from T_0 to T_N. As stated, this result does not even suggest that time fails to have a direction at all. Indeed, the necessity of invoking the time reversal operation on instantaneous states suggests just the opposite: it suggests that even for an instantaneous state, there is a fact about *how it is oriented with respect to the direction of time*.

Given certain facts about the time reversal operator (in particular, given that particle positions or field values do not change under the operation), Time Reversal Invariance, as stated above, is still a very important feature of physical laws. For, as Albert insists, this sort of Time Reversal Invariance implies that *in a certain sense* anything that happens in the universe can, as a matter of physical principle, happen 'in reverse': if ice cubes can melt into puddles, then puddles can spontaneously freeze back into ice cubes, for example. But notice that this sense of 'in reverse', far from delegitimizing the passage of time, presupposes it: the melting of an ice cube 'in reverse' requires that the puddle stage *precede* the ice cube stage. So to get from Time Reversal Invariance of the laws (in this sense) to the objective *absence* of a direction of time would require some extensive philosophical argumentation in any case.

Finally, let's even set Albert's objections aside. That is, let's suppose that we can understand the time reversal operation without there being an objective direction with respect to which the reversal occurs, let's suppose that the fundamental laws of nature make no distinction between the past direction and the future direction of time. Would it even now follow unproblematically that, as far as physics is concerned, there is no direction of time, no distinction between the past-to-future and future-to-past directions?

---

[9] One has to be clear about what it means for a state to be instantaneous: instantaneous states, for example, do not include *velocities* of particles, or, more generally, the rate-of-change-with-time of any quantity. See Albert's discussion.

If *all* that physics ever appealed to in providing accounts of physical phenomena were the dynamical laws, then we would seem to have a straightforward Ockham's Razor argument here. But the laws of nature *alone* suffice to explain almost nothing. After all, it is consistent with the laws of physics that the universe be a vacuum, or pass from the Big Bang to the Big Crunch in less time than it takes to form stars. The models of fundamental physical law are infinitely varied, and the only facts that those laws *alone* could account for are facts shared in common by all the models. In all practical cases, we explain things physically not merely by invoking the laws, but also by invoking *boundary conditions*. So to argue that the direction of time plays no ineliminable role in physics demands not only a demonstration that no such direction is needed to state the laws, but that no such direction plays any role in our treatment of boundary conditions either. But this latter is far from obvious. We will return to this point towards the end of this Chapter.

3. EPISTEMIC OBJECTIONS

Let's return for the moment to the violation of CP invariance displayed in neutral kaon decay. We noted above that this phenomenon seems to imply that the laws of nature are not Time Reversal Invariant in any sense, and hence that the laws themselves require an intrinsic asymmetry in time directions, and hence that space-time itself, in order to support such laws, must come equipped with an orientation. But one might still object that this orientation (let's call it Kaon Orientation) has nothing to do with the supposed large-scale and manifest asymmetry involved in the passage of time. For it is only in rather special and recondite circumstances that Kaon Orientation manifests itself, while the passage of time is evident and ubiquitous. More directly, even if there is an intrinsic Kaon Orientation to space-time, in most normal circumstances we would not be in a position to determine what it is.

We can call this sort of objection an epistemic objection: it does not directly deny the logical or physical acceptability of the existence of a fundamental temporal asymmetry, but insists that even if such an asymmetry is postulated, we would not be able to tell what it is. In the case of the Kaon Orientation, the objection is merely practical: kaon decay experiments (or other physical phenomena whose outcomes depend on the Kaon Orientation) are relatively rare and inaccessible. But the epistemic objection can be raised in an even more thoroughgoing and radical manner, in a form which, if valid, would make the passage of time epistemically opaque even if physically real.

Price raises the objection in one way, D. C. Williams in another. Let's begin with Price's formulation:

> In practice, the most influential argument in favor of the objective present and the objective flow of time rests on an appeal to psychology—to our own experience of time. It seems to us that time flows, the argument runs, and surely the most reasonable explanation of this is that there is some genuine movement of time which we experience, or in which we partake.
>
> Arguments of this kind need to be treated with caution, however. After all, how would things seem if time didn't flow? If we suppose for the moment that there is an objective flow of time, we seem to be able to imagine a world which would be just like ours, except that it would be a four-dimensional block universe rather than a three-dimensional dynamic one. It is easy to see how to map events-at-times in the dynamic universe onto events-at-temporal-locations in the block universe. Among other things, our individual mental states get mapped over, moment by moment. But then surely our copies in the block universe would have the same experiences that we do—in which case they are not distinctive of a dynamical universe after all. Things would seem this way, even if we ourselves were elements of a block universe. (Price 1996, pp. 14–15)

The short diagnosis of the foregoing is that it is an argument by 'surely': Price simply asserts that the crucial conclusion follows from his premises even though it is by no means evident that it does. The point can perhaps be made more clearly if we switch to the form that Williams uses.

That form begins by granting that there is a direction of time, so the past-to-future direction differs from the future-to-past direction. But now the observation is made that we also accept the Time Reversal Invariance (let's set aside the kaon interactions as irrelevant here) as stated above: for any physically possible sequence of states $T_0, T_1, \ldots T_N$ running from past to future, there is a physically possible sequence $T_N{}^*, \ldots T_1{}^*, T_0{}^*$ running from past to future. For example, given the actual sequence of physical states of your body over the last ten minutes, the time-reversed sequence of time-reversed states is also physically possible. Somewhere on some other planet (as far as the laws of physics go) some such sequence could exist, unproblematically time reversed relative to the sequence of states that make you up. Let's call this sequence of states your *time-reversed Doppelgänger*. But, the objection goes, there is an obvious one-to-one mapping from the Doppelgänger's states to yours. So the Doppelgänger would *surely* have qualitatively identical experiences to yours, only with the whole process oppositely oriented in time. Here is Williams's description:

> It is conceivable too then that a human life be twisted, not 90° but 180°, from the normal temporal grain of the world. F. Scott Fitzgerald tells the story of Benjamin Button who was born in the last stages of senility and got younger all his life till he dies a dwindling embryo. Fitzgerald imagines the reversal to be so imperfect that Benjamin's stream of consciousness ran, not backward with his body's gross development, but in the common clockwise manner. We might better conceive a reversal of every cell twitch and electron whirl, and hence suppose that he experienced his own life stages in the same order as we do ours, but that he observed everything around him moving backward from the grave to the cradle.[10] (Williams 1951, p. 113)

If we accept that the relevant physics is Time Reversal Invariant, then we accept that your time-reversed Doppelgänger is physically possible. Let's suppose, then, that such a Doppelgänger exists somewhere in the universe. What should we conclude about its mental life?

The objector, of course, wants to conclude that the mental state of the Doppelgänger is, from a subjective viewpoint, just like ours. So just as we judge the 'direction of the passage to time' to go from our infant stage to our gray-haired, so too with the Doppelgänger. But that direction, for the Doppelgänger, is oppositely oriented to ours. So the Doppelgänger will judge that the temporal direction into the future points opposite to the way we judge it. And if we insist that there is a direction of time, and we know what

---

[10] This particular conceit, that a time-reversed person would *see* the world around him as if it were a movie shown backwards, is not supported by any physical considerations of time reversibility. Presumably, if your time-reversed Doppelgänger experiences anything at all, it will be a subjectively indistinguishable experience from your present one: the experience of things happening in their accustomed order. The question of whether it would even be physically possible for the Doppelgänger to interact with a non-time-reversed environment (or, equivalently, whether we could physically interact with a time-reversed environment which contained puddles constituting themselves into ice cubes) to allow for *perception* of that environment is a tricky technical question. The problem arises since the entropy gradient in the environment has to be oppositely oriented from the entropy gradient in the observer, and it is not clear whether these can be made to mesh, or, if they do, whether the meshing condition produces what would normally be called perception. Perception typically suggests a wide array of counterfactual dependence so that the mental state of the observer would track the state of the environment over a range of possible variations. The entropy meshing might require that the brain state of the observer and the state of the environment be exquisitely fine-tuned to each other, preventing such counterfactual resilience. In any case, the brain state of someone watching a movie being run backward is not the time reverse of the brain state of someone watching the original sequence of events, and the time reverse of the optical array produced by a melting ice cube is not the same as the optical array produced by the backward-running movie of the melting, so simple time reversal arguments cannot establish under what conditions one would be able to see a time-reversed melting ice cube.

it is, then we must say that the Doppelgänger is deceived, and has mistaken the direction of time. But now we become worried: the Doppelgänger seems to have exactly the same *evidence* about the direction of time as we do. So how do we know that (as it were) *we* are not the Doppelgängers, that *we* are not mistaken about the direction of time? If there *is* a direction of time, it would seem to become epistemically inaccessible. And at this point, it seems best to drop the idea of such a direction altogether. But is this correct?

In order to facilitate the discussion, I will refer to corresponding bits of the Doppelgänger with a simple modification of the terms for parts of the original person. For example, I will speak of the Doppelgänger's neuron*s: these are just the bits of the Doppelgänger that correspond, under the obvious mapping, to the original's neurons. We can unproblematically say that the Doppelgänger's neuron*s fire*, meaning that they undergo the time reverse of the process of a normal neuron firing. It *may* be that neuron*s can be properly called neurons and the firing* may be properly called firing, but we do not want to presuppose that at the outset, so one must remain alert to the asterisks.

So the first question is: given the physical description of the Doppelgänger that we have, what can we conclude about its mental state? The answer, I think, is that we would have no reason whatsoever to believe that the Doppelgänger has a mental state at all. After all, the physical processes going on in the Doppelgänger's brain* are quite unlike the processes going on in a normal brain. Nerve impulse*s do not travel along dendrites to the cell body, which then fires a pulse out along the axon. Rather, pulses travel up the axon* to the cell body*, which (in a rather unpredictable way) sends pulses out along the dendrite*s. The visual system* of the Doppelgänger is also quite unusual: rather than absorbing light from the environment, the retina*s emit light out into the environment. (The emitted light is correlated with the environment in a way that would seem miraculous if we did not know how the physical state of the Doppelgänger was fixed: by time-reversing a normal person.) There is no reason to belabor the point: in every detail, the physical processes going on in the Doppelgänger are completely unlike any physical processes we have ever encountered or studied in a laboratory, quite unlike any biological processes we have ever met. We have *no reason whatsoever* to suppose that any mental state at all would be associated with the physical processes in the Doppelgänger. Given that the Doppelgänger anti-metabolizes, etc., it is doubtful that it could even properly be called a living organism (rather than a living* organism*), much less a conscious living organism.

Now the response is likely to chime in: the Doppelgänger's physical state is not unfamiliar: it is *just like ours* save for the direction of time. That is so: but the difference is no minor one. It turns emission into absorption, normal metabolic processes into weird and unexampled anti-thermodynamic ones.

No, no, the response insists: since the gradient of the entropy (or whatever) has been reversed, *the direction of time itself has been reversed*, and, oriented to the thus-defined direction of time, the physical processes are just like normal ones. This response, of course, has a name: *petitio principii*. The aim of the argument it to *show* that there is no intrinsic direction to time, but only, say, an entropy gradient. But it achieves its aim only if we are convinced that the Doppelgänger has a mental state 'just like ours', and the only way to make that claim even vaguely plausible is to assert that the Doppelgänger's *physical* state is not, in any significant sense, time reversed (relative to any physically significant direction of time) at all. And that is precisely to beg the question.

Having worked through the response to Williams, the response to Price is even starker. He imagines a Doppelgänger that is not just reversed in time, but a Doppelgänger in a world *with no passage of time at all*, i.e. (according to his opponent) in a world in which there is no time at all, perhaps a purely spatial four-dimensional world. So it is not just that the nerve pulse*s of this Doppelgänger go the wrong way (compared to normal nerve pulses), these nerve pulse*s don't go anywhere at all. *Nothing* happens in this world. True, there is a mapping from bits of this world to bits of our own, but (unless one already has begged the central question) the state of this world is so unlike the physical state of anything in our universe that to suppose that there are mental states at all is completely unfounded. (Even pure functionalists, who suppose that mental states can supervene on all manner of physical substrate, use temporal notions in defining the relevant functional characterizations. Even pure functionalists would discern no mental states here.)

All the participants in this debate accept the supervenience of the mental on the physical. If you believe in the passing of time, then that is among the relevant physical characteristics of the subvenience base. To simply suppose that it is not, that the character of mental states cannot possibly depend on how time is passing for the brain states, and to disguise this critical assumption with no more than an offhand 'surely', is to miss the nature of the dispute altogether.

Up to now, we have been considering only the following question: given that we regard the time-reversed Doppelgänger as physically possible, what should we conclude about the mental state (if any) that such a Doppelgänger would have. There is, however, another slightly different argument in the neighborhood that needs to be carefully distinguished from this one. This argument goes as follows: suppose you were to *actually find* or have *strong evidence for the existence of* some such Doppelgänger (on a distant planet, say), then would that not change your view about the objective passing of time?

This is quite a different matter. It is perhaps too flippantly posed: it might not even be *physically possible* to 'find' (i.e. observe or encounter) such a Doppelgänger if there were one (see footnote 5 above): the *light* that would enter our eyes from such a Doppelgänger would not, for example, be correlated with the shape or movements of the Doppelgänger, and it is unclear what other sort of evidence could convince us of its existence. But if we *were* somehow convinced of its existence, what would we conclude?

If we stick to the idea that time passes, and that the Doppelgänger really is a time-reversed copy of a normal person, then, as argued above, we would have no reason to suppose it conscious, or to speculate on the content of its mental states if it were conscious. But quite apart from its mental states, the Doppelgänger would behave in a radically anti-thermodynamic way (relative to our time direction), and there would be correlations between the Doppelgänger and its environment which we would judge, although physically possible, to be wildly unlikely and conspiratorial. That is, physics as we presently practice it would regard the Doppelgänger as a physical possibility of a sort almost unmeasurably unlikely ever to occur. So if we were to actually *find* or to *come to believe in the reality* of such a thing, we would have very good reason to *revisit* our physical theory, including any views we might have about the passage of time. Perhaps (but who knows?) I would abandon my views about the passage of time in favor of a different view, according to which the existence of such a Doppelgänger is not so unlikely. But just because I regard the Doppelgänger as physically possible, and think that the existence of such a Doppelgänger would provide good reason to revise my views on the nature of time, it does not follow that I have reason to regard my views *now* (in the absence of evidence of a Doppelgänger) as impeached. Nor is this a good inference: I regard the Doppelgänger as physically possible, and I would regard a Doppelgänger as good reason to reject an absolute passage of time, therefore I regard the absence of absolute passage as *physically possible*. If there is passage, a Doppelgänger may be possible—a physically possible but *evidentially misleading* object. An analog: even given a purely local stochastic physics, it is possible to observe violations of Bell's inequality in a laboratory (because of an objectively *very unlikely* set of outcomes). Were those outcomes to occur, it would be rational to reject the purely local physics. But it does not follow that the purely local physics underwrites the *physical possibility* of non-locality, just the physical possibility of evidence that rationally leads us incorrectly to accept non-locality.

So none of the arguments for the epistemic inaccessibility of the direction of the passage of time goes through without already begging the question at hand. And having exhausted the logical, the scientific, and the epistemic arguments, there are no stock arguments left. The usual philosophical arguments, which have induced Price and Williams and many others to reject an objective passage of time, have no force whatsoever.[11]

4. THE CASE IN FAVOR OF THE PASSING OF TIME

The failure of the stock arguments against the passing of time does not *per se* constitute a positive argument in favor of the claim that time passes. What, then, can be said in defense of the claim that time, in its own nature, passes?

I do not have much that is original to offer here, so presenting the positive case is mostly an organizational task. It will be helpful in presenting the case to divide the opponents of the passing of time into two camps, and to deal with them in turn. We may denominate the two camps the *conciliatory* opponents and the *in-your-face* opponents.

Gödel, as cited above, is an archetypal in-your-face opponent. The absence of an objective passing of time, according to Gödel, means that Parmenides, Kant, and the modern idealists were right when they considered 'change as an illusion or an appearance due to our special mode of perception'. Physical

---

[11] There is one famous source of arguments against the passage of time that I have not mentioned: that of J. M. E. McTaggart (1927, chapter 33). The reason is that this argument is a mare's nest of confusions, the proper use of which is to give to students with instructions like 'Find the fallacy of composition' and 'Explain why McTaggart's response to Russell is question-begging'. But since the terminology *A series theory of time* and *B series theory of time* have crept into the literature, let me make some comments. The theory of time's passage I defend focuses on the *B* series: all events are ordered by a transitive, asymmetrical relation of earlier and later. Given events ordered in a *B* series, one can define an infinitude of *different A* series that correspond to taking different events as 'now' or 'present'. McTaggart's argument is marred throughout by his use of the phrase 'the *A* series' when there are, in fact, an infinitude of such. Any theory that denies a fundamental asymmetric relation of earlier than (or later than), and hence denies an intrinsic direction of time, ought not to be called a *B* series theory but rather a *C* series theory. So I am not arguing for an *A* series theory over a *B* series theory, I am arguing for a *B* series theory over a *C* series theory, and get the many *A* series for free.

reality, as it is apart from conscious beings, contains no change at all. Rivers, in themselves and apart from our representations of them, do not flow; stars, in themselves and apart from our representations of them, do not collapse; atoms do not radiate; and so on. Change is an illusion in the same way that the apparent comparative size of the moon at different positions in the sky is an illusion: not even an optical trick, like the apparent bending of a stick in water (which has an explanation in terms of the behavior of light), but a product solely of the cognitive machinery of humans. According to the in-your-face opponent, outside of human (or other conscious) beings, there is nothing that corresponds to change at all. And in a world without change, there is surely no need for the passage of time.

The response to the in-your-face opponent is straightforward. We ought to believe that there is objective change because (1) the world is everywhere and always presented to us as a world in which things change, (2) change appears to affect all physical things and (unlike, say, pain) not to be a feature of only a small, select group of objects, and (3) there are no good arguments to suggest that this change is illusory. That is, the basic approach to ontology is always to start with the world as given (the manifest image), and then adjust or adapt or modify or complicate that picture as needed, under the pressure of argument and observation. But neither argument nor observation suggests that things do not change in the mind-independent world, so we ought to think that they do.

Another way to put the response is this: why should we think that the physical world has temporal extension (apart from the 'direction of time') at all? Why think that it has a spatial aspect, or spatial extension? Kant, of course, denied these as well, as did Parmenides and the modern idealists. But no contemporary 'analytical' philosopher, people like Price or Horwich or Sklar or Albert or Williams or Earman—in short, no one who takes modern physics as the touchstone for ontology—denies that the physical world, in itself, has spatio-temporal extension. But why believe *that*? Presumably because that is how the world is, in the first place, presented to us and because our further investigations—both logical and scientific—have not given us any grounds to question it. So too for the reality of change.

The conciliatory opponent, on the other hand, does not claim that change is mere illusion or appearance. The conciliatory opponent is happy to say that things change, that rivers flow, that stars collapse, and that their doing so does not depend in any way on human modes of perception or consciousness. The conciliatory opponent simply wants to insist that such objective, mind-independent change does not require that there be any objective, mind-independent, *intrinsic* passage of time. The attempt to analyze change without the passage of time proceeds in two steps.

The first step is exemplified in this passage from Williams:

> Motion is already defined and explained in the dimensional manifold as consisting of the presence of the same individual in different places at different times. It consists of bends or quirks in the world line, or the space-time worm, which is the four-dimensional totality of the individual's existence. (Williams 1951, pp. 104–5)

According to Williams, the motion of the asteroid from Earth to Mars is just a matter of the asteroid being differently situated with respect to those planets at different times or, to take an easier case of change, my losing weight is just a matter of my space-time worm being thicker at one end than at another. Since these are objective, mind-independent facts about space-time worms, the changes are equally objective and mind-independent.

The rub, of course, is that the asteroid being differently situated at different times is consistent both with a motion from Earth to Mars and with a motion from Mars to Earth, and my space-time worm being thicker at one end than at another is as likely to indicate my gaining weight as my losing it. Motions and changes are not merely a matter of things being different at different times, but also critically a matter of which of these times are *earlier* and which *later*. To preserve a form of 'change' that cannot distinguish losing weight from gaining it is not to preserve change at all, so this attempt at reconciliation fails.

The conciliatory opponent is not done. The next step is to admit that some objective correlate to *earlier* and *later* must be found, so weight loss can be distinguished from weight gain, but to deny that this 'direction of time' need be a matter of the passage of time itself. An arrow must be provided, but the resources exist without adding anything intrinsic to the (undirected) spatio-temporal structure. Rather, the direction from earlier to later is nothing but, e.g., the direction of increasing entropy. You are losing weight if the universe at the thin end of your space-time worm has a higher entropy than the universe at the thick end, and the asteroid is traveling from Earth to Mars if the entropy at the Mars end of the trip is higher than the entropy at the Earth end. If there is no difference in the entropy (e.g. if the universe is in thermal equilibrium), then there is no longer a distinction between Earth-to-Mars and Mars-to-Earth trips. But we have never had any acquaintance with situations in complete thermal equilibrium, and there may never be such situations, and even if there were, we could not experience them (since our own operation demands an entropy gradient), so the account is adequate to everything we ever have experienced or will experience.

This sort of conciliatory opponent seems to have Ockham's Razor as a weapon. After all, the Second Law of Thermodynamics says that entropy never decreases. As stated, this appears to be a contingent rule relating the entropy gradient to the direction of time (as implicit in the word 'decrease'). But why not turn the 'law' instead into an implicit *definition* of the direction of time: entropy never decreases because if the entropy gradient is not zero, then the forward direction of time *just is* the direction in which entropy gets larger! Our fundamental ontology is reduced (we no longer have *both* an entropy gradient *and* a direction of time), and the resulting definition of the direction of time is materially adequate, since we *do* think that entropy never decreases.

There are, however, several objections to this procedure. First, the reinterpreted 'Second Law' no longer has the *content* of the original. The original supposed that there is a direction of time—represented by an orientation on the space-time manifold—and that entropy never decreases *relative to that orientation*. That is, the original implied that the direction of entropy increase would always be the same, that entropy would ever increase (or at least not decrease) in a given, globally defined, direction. The new 'Law', as an implicit definition, puts no constraints on the shape of the global entropy curve at all. Entropy could go up and down like the stock market, but since the 'direction of time' would obligingly flip along with the entropy changes, entropy would still never decrease. And since we *don't* think that the entropy curve fluctuates in this way, the original has explanatory power that the reinterpreted version lacks.

(There is an interesting parallel here to those practitioners of analytical mechanics who want to take $\mathbf{F} = m\mathbf{A}$ to be a definition of force rather than a law of nature, as Newton had it. This too reduces one's ontology, but at the price of seriously distorting the content of the scientific theory. Newton certainly did not take his law to be a definition, and nor do we—since we no longer accept it. Further, under some very mild assumptions (e.g. that the masses of bricks do not change when different springs are attached to them, and the forces produced by springs do not change when they are attached to different bricks) Newton's Second Law has testable implications, and could be experimentally refuted.)

The situation gets even worse once thermodynamics is subsumed under statistical mechanics, for then the Second Law is no longer taken to be nomically guaranteed. Global entropy *could* decrease, and even has a calculable chance of doing so, and certainly will do so if we wait long enough; and local entropy, due to random fluctuations, decreases (if briefly) all the time. So the original Ockham's Razor argument—since the anisotropy of entropy increase and the anisotropy of the passage of time always coincide, why not be economical and identify them?—loses all of its force.

But let's put all that aside. Let's grant, for the moment, that entropy strictly increases monotonically throughout the history of the universe. And let's further grant (setting aside actual physics) that the fundamental laws of nature are completely Time Reversal Invariant, and can be stated without reference to any directionality in time. Would all that suffice to rationally compel us to *identify* the passage of time with the increase in entropy?

I do not think so. For simplicity, let's assume that the Time Reversal Invariant laws are deterministic, like Newtonian mechanics or classical electromagnetic theory. Then, as noted above, the total state of the world is accounted for not merely by reference to the laws, but by reference to the laws and the boundary conditions (let's not yet say *initial* conditions) of the universe. And the mathematical character of these boundary conditions, relative to the laws, is very particular. Let's assume that there are boundary conditions in the past (the Big Bang), in the future (the Big Crunch), and at the spatial limits (if the universe is not spatially closed). Now one is not free to arbitrarily specify boundary conditions along *all* of these boundaries and then to expect there to be a solution that respects the laws everywhere in the interior. Rather (leaving the problem of the spatial boundaries out for a moment), one is (relatively[12]) free to specify whatever conditions one likes on one of the boundaries, and then there will typically be a unique global solution to the laws that determines the conditions on the other boundary. And of course, when providing explanations and accounts of things, what we actually do is specify the state on the *initial* (i.e. earliest) boundary, and regard the state in the interior and on the final boundary to be *explained* or *produced* from the initial conditions and the operation of the laws through time.

The world might not have been this way. We could imagine deterministic laws of such a form that one could freely specify boundary conditions on all the boundaries of a system and be assured of a unique global solution consistent

---

[12] There may be certain constraints on the initial boundary conditions, having to do with charges and divergences of fields and so on, but these are of a different character than the problem we are presently discussing.

with those conditions. It is a stunning, and contingent, fact that the laws—or the best approximations to the laws that we have been able to discover so far—have the mathematical form just described. And all this holds *even if the laws themselves are completely time reversible in whatever sense one might choose.*

This particular form of the laws of nature leads us into an asymmetrical treatment of the boundary conditions of the universe. It seems that given one set of boundary conditions together with the laws of nature we are able to adequately account for the other set of boundary conditions as well as the complete contents of the universe. Among the contents so explained is, of course, the global increase of entropy (insofar as it holds). So we have the following situation: if the asymmetrical treatment of the 'initial' and 'final' boundary conditions of the universe is a reflection of the fact that time passes *from* the initial *to* the final, then the entropy gradient, instead of explaining the direction of time, is explained by it.

The asymmetrical treatment of the boundary conditions of the universe is well known: we think that the universe started off in a macroscopically atypical but microscopically typical state (relative to a natural measure on the phase space of the universe), and that it will end in a macroscopically typical (or more typical) but microscopically atypical state. The 'initial' state is macroscopically atypical in that it has extraordinarily low entropy, but microscopically typical in that temporal evolution from it leads to higher entropy. The 'final' state is macroscopically typical, in that it has high entropy, but microscopically atypical in that temporal evolution from it ('backwards in time' as we would say) leads immediately and monotonically to lower entropy. Now one could try to deny that this sort of asymmetry in boundary conditions actually obtains (Price approvingly flirts with the idea of a 'Gold universe' which has low entropy at both ends), but every result we have from empirical enquiry suggests that this is how the universe is.

Since we cannot explain any of the actual contents of the universe by reference to the laws alone, we either have to abandon any such explanation or try to bring boundary conditions into such an explanation. But, as we have seen, if we are to posit constraints on boundary conditions, we had best not apply those constraints to *all* the boundaries, since it is typically not the case that the laws are consistent with constraints on all the boundaries. Rather, we seem to need only constraints on *one* boundary, together with the laws, to get us the universe. So the only question left is: which boundary should be subject to constraints, and what, if anything, does the selection of that boundary suggest about the passage of time?

One might think that although the two boundaries are treated differently, there is still a sort of explanatory symmetry between them. That is: postulate a macroscopically atypical but microscopically typical state, plus the laws, and one can explain the macroscopically typical but microscopically atypical state from them: the latter was generated from the former by means of the operation of the laws. But equally: postulate a macroscopically typical but microscopically atypical state at one end, plus the laws, and one can 'generate' a macroscopically atypical but microscopically typical state from them. Pick one end, add the laws, and you can explain the other end: which end you pick as *explanans* and which as *explanandum* is up to you.

But this supposed symmetry is illusory. The problem is this: in order to account for the universe as we see it, we need more than the laws: we need a constraint on one of the boundaries. That constraint, together with the operations of the laws, then suffices to account for the nature of the other boundary. But in order for this to work *the constraint must itself be specifiable independently of what will result from the operation of the laws*. A homely example: we want to explain the occurrence of an earthquake in terms of the preceding state of the underlying geology and the dynamics that governs them. This works fine so long as the precedent state is described in terms like pressures and fissures and plate movements and so on, terms that can be plugged into the laws to determine how the system will evolve. It does *not* work as an explanation if the only characterization one provides of the underlying geology is 'in a state such as to lead to an earthquake in the near future'. Explaining the occurrence of the earthquake in these terms is clearly empty.

But exactly this sort of thing occurs if we take one of the boundary constraints as the basis (together with the laws) of the explanation of the other. For despite the surface similarity, the terms 'macroscopically atypical' and 'microscopically atypical' as used above have a completely different logic.

The initial state of the universe is macroscopically atypical in a way that can be completely characterized without any mention of the details of its dynamical evolution. The initial macrostate is atypical because it has low entropy, i.e. because it occupies (relative to the natural measure) a very very very small volume of phase space. One can characterize this atypicality without any mention at all of how such a state will evolve. The final state, however, is microscopically atypical in a way that can *only be characterized in terms of how the state will 'evolve' though time*. It is microscopically atypical because temporal evolution in one direction from it will lead, over a very long period of time, to monotonically lower entropy. A generic microstate (relative to the natural measure over phase space) leads to higher (or constant) entropy whichever way one time evolves from it in accordance with the laws. The initial state is unusual in what it *is*, while the final state is unusual only in what it will *become* (in a generic sense of 'become' which covers backward evolution).

So if we want to explain the world as we see it—including the ubiquitous entropy gradient around us—this seems like a viable strategy: postulate in addition to the dynamical laws of nature a logically independent constraint on the initial state of the universe—that it has low entropy—and nothing else. On the assumption (and it is a big assumption, but one that we constantly make) that typical conditions (relative to the natural measure) require no further explanation, we would then expect a world of constantly increasing entropy, as we see. The world should evolve towards ever higher entropy, with the states always retaining the following peculiarity: their time reverses lead to lower, rather than higher entropy. But the 'peculiarity' is completely accounted for: time reverses of the states lead to lower entropy because the states themselves arose out of lower entropy. In particular, the microscopic atypicality of the final state is completely accounted for by *how it was generated or produced*.

But we cannot run this trick in reverse. Even though the laws themselves might run perfectly well in reverse, even though the time reverse of the final state might give rise to the time reverse of the initial state, we cannot specify an independent, generic constraint on the final state that will yield (granting the final macrostate is typical) ever decreasing entropy in one direction. The 'atypical' microstates are miscellaneous and scattered: they have *nothing* in common that can be described independently of the detailed temporal evolution to which they give rise. (Thus: a slight modification of the dynamical laws would lead to essentially no change in which initial states are macroscopically atypical, in that they have low entropy, but would completely alter the set of atypical high-entropy states whose time evolution in either direction leads to low entropy.) This asymmetry in the way that the initial and final boundary states are atypical forces an asymmetry in the explanatory scheme. And the explanatory scheme then uses the direction of time in accounting for the universe, even when using time-symmetric laws. The atypical final state is accounted for as the *product of an evolution from a generically characterized initial state* in a way that the initial state cannot be explained as a product of evolution from a generically characterized final state.

The basic structure of the explanatory situation can be seen as follows. On the one hand, this sort of explanation makes essential use of a notion of *typicality*: having granted that the state on one boundary is low entropy, we show that the vast majority of microstates compatible with that macrostate lead to higher entropy. This is taken to provide an *explanation* of the entropy increase in that evolution. And it appears that we can repeat the strategy: the state of the universe after a billion years is still low entropy, so we can expect the entropy to continue to rise. But when considering the state after a billion years, a puzzle arises: not only is entropy increase typical for evolution in the direction away from the low-entropy boundary, entropy increase is typical—in exactly the same sense—for evolution *towards* the low-entropy boundary. So either the typicality arguments do not really give explanations, or the expectations that arise from the typicality analysis must somehow be *trumped* when projecting backward in time. For any but the initial state, typical behavior is actual behavior only in one direction of time: behavior in the other direction is (in the *mathematically defined sense*) always atypical. That is, when looking backward in time, 'typical' behavior—behavior exhibited by almost all microstates compatible with the macrostate—is not *usual*: in fact, it never occurs!

If we are to maintain that typicality arguments have any explanatory force—and it is very hard to see how we can do without them—then there must be some account of why they work only in one temporal direction. Why are microstates, except at the initial time, always *atypical* with respect to backward temporal evolution? And it seems to me we *have* such an explanation: these other microstates are *products of a certain evolution*, an evolution guaranteed (given how it *started*) to produce exactly this sort of atypicality. This sort of explanation requires that there be a fact about which states produce which. That is provided by a direction of time: earlier states produce later ones. Absent such a direction, there is no account of one global state being a cause and another an effect, and so no account of which evolutions from states should be expected to be atypical and typical in which directions. If one only gets the direction of causation from the distribution of matter in the space-time, but needs the direction of causation to distinguish when appeals to typicality are and are not acceptable, then I don't see how one could *appeal* to typicality considerations to *explain* the distribution of matter, which is what we want to do.

So even apart from the (actual) lack of time reversal invariance of the laws of nature, we would have reason to accept an intrinsic asymmetry in time itself: an asymmetry that plays a role in explaining both the nature of the final state of the universe and the constant increase in entropy that connects the initial to the final state.

Above and beyond and before all of these considerations, of course, is the manifest fact that the world is given to us as changing, and time as passing, and everyone takes for granted that their situation is importantly different before and after a visit to the dentist (even if they are at equal temporal removes), and importantly different toward the beginning and end of their lives (even though there may, in each case, be a long stretch of life lying to one side or the other) and that all the philosophizing in the world will not convince us that these facts are mere illusions. Even Descartes at his most skeptical, willing to question the existence of the external world, never questions the passage of time. Nor is there any evidential reason to suggest that such passage is 'only in our minds': it is the passage of time (in the right conditions) that leads iron to rust and fires to burn down. But to insist on these observations is somehow not to engage in philosophy, so I will desist.

5. HOW SHOULD THE PASSAGE OF TIME BE REPRESENTED MATHEMATICALLY?

If we accept that it is an intrinsic, objective feature of time that it passes (or that there is an intrinsic, objective, distinction between future-directed timelike vectors and past-directed vectors), then that feature ought to be incorporated into our mathematical representation of space-time. Granting that a Relativistic metric already represents objective spatio-temporal structure, but that such a metric includes no intrinsic asymmetry between the timelike directions, what needs to be added to the math?

As already mentioned above, all that seems to be required, from a mathematical point of view, is an orientation: an indication of which of the two globally definable timelike directions is to the future and which is to the past. Furthermore, as has also been mentioned, the violation of CP invariance observed in nature already appears to demand an orientation, which we have called the Kaon Orientation (or, more generically, the Weak Interaction Orientation). But now the Ockhamist siren song may be heard: why add *two* orientations to our ontology? Why not assume that the Kaon Orientation *just is* the passage-of-time orientation? Such an assumption would reduce our ontology at no cost to empirical adequacy.

The identification of the two orientations appears not to be a mere matter of semantics: if the orientations are ontologically distinct, then on the surface it seems that they could have been oppositely oriented to the way they are, while if there is only one, they could not. But this is a rather intricate question, since the violation of CP already requires specification of the spatial parity orientation. Let's leave these complications aside.

But there is another worry about identifying the two orientations. The passage of time is ubiquitous and manifest. Violation of CP is rare and subtle. If the orientation of time's passage were just the orientation that the Weak Interaction is attuned to, wouldn't we have trouble explaining how we can be so surely and easily aware of the direction of the passage of time?

A possible analogy: we naively begin with the idea that there is a manifest intrinsic anisotropy of *space*, which we denominate 'up' and 'down'. There then ensues a debate: is the distinction reflective of some innate structure of *space itself*, or merely a matter of how some material substance is distributed in an intrinsically isotropic space. The Newtonian takes the latter approach: 'up' and 'down' are determined not by space itself, but by the gradient of the gravitational potential. But notice: for this account to be acceptable, it had better be that the phenomena around us, and indeed our own bodies, are strongly coupled to the gravitational field. For we can easily distinguish up from down in everyday life, and if we ever become dizzy or disoriented (as occasionally happens) we can quickly regain our bearings. The explanation for this latter ability depends on the effect of gravitation on the semicircular canals in the inner ear. If the direction of time were only manifest in things like kaon decay, wouldn't we need a similar (and much more sophisticated) physical organ in the brain to orient ourselves to the passage of time?

But once we have proposed the analogy, its absurdity becomes manifest. We can indeed become disoriented and lose track of the directionality of space (whether this directionality is intrinsic to space itself or due to its contents is immaterial). But what would it even mean to be 'disoriented with respect to time', not to be sure, as it were, which way time is passing? This is certainly not a familiar or imaginable psychological state. And what would it mean to have a device (like the semicircular canals) that could serve to 'orient us in time'? After all, the *definition* of a good detector of something (whether it be gravitational fields or magnetic fields or cosmic rays) is already given in time-oriented terms: a good magnetic field detector (like a compass) is a device that, if *started* in a ready state (well oiled, level, and with the needle pointing in any direction at all), will *end* in a state in which some internal pointer (the needle) is pointing in the direction of the local magnetic field. To say that a compass is working as a good indicator of the local magnetic field requires knowing the direction of time: the time reverse of the behavior of a good compass (needle going from an arbitrary direction to parallel with the local field) is the behavior of a bad magnetic field detector (going from pointing in the direction of the field to pointing in an arbitrary direction). Furthermore, *reacting* to the output of such a device and *basing* our further behavior on that outcome (e.g. deciding which way to go) are time-oriented behaviors, so if we could react to the reading of a 'time orientation' device and adjust our behavior to it, we would already have to be functioning in a time-oriented way.

Of course, this situation is perfectly general—it has nothing to do with any particular account of the direction of time. If one believes that the direction of time is just the direction of the local entropy gradient, still one does not think that we 'orient ourselves in time' by measuring the entropy at different times and figuring out in which direction it is increasing (not least because such measuring and figuring out already require increasing entropy). Entropy, after all, is not a first-order physical magnitude (like an electric field) for which one could even build a measuring apparatus. Rather, anyone who thinks that time direction is the direction of increasing entropy will say that we experience time as passing in a certain direction *simply because our entropy is increasing in that direction*, without the need for any further apparatus to determine the increase. In short, we experience the direction of time in a certain way because we simply are, as physical objects, time directed (even if 'time direction' is a matter of entropy gradient). The chain of observing-one-thing-by-means-of-another must come to an end somewhere: there must be elements of our experience that are determined directly by our physical constitution rather than by mediation through detectors.

So if we experience the passage of time simply because we are (as physical objects) things in time, and time passes, then the other ways that the orientation of time appears in physics are not important. The rarity of phenomena like kaon decay would not stand in the way of identifying the Kaon Orientation with the orientation of the passage of time. But since I am suspicious of the Ockhamist position that such identifications are always desirable *per se* I find myself simply agnostic about whether the orientations ought to be identified.

There is one final issue that needs to be addressed. I have claimed that the only mathematical gadget one needs to add to represent the passage of time is an orientation. But surely, it will be urged, this is hardly sufficient to adequately represent the metaphysical nature of passage. After all, one might have to add some intrinsic, anisotropic structure to space (a structure that would induce a partial ordering of events), but that would surely not justify the claim that space 'passes'. Similarly, the fan of entropy increase can admit violation of CP invariance, and hence the need for a physical orientation in space-time, without thereby being forced to admit that the orientation has anything to do with the problem of 'the passage of time'. So isn't the mathematical structure used to represent the passage of time inadequate to its representational task; doesn't it leave the 'intrinsic nature of time's passage' completely obscure?

Yes, it does. There is nothing intrinsic about an orientation that makes it particularly suited to represent the passage of time (as opposed to, say, the Kaon Orientation, if it is distinct), just as there is nothing intrinsic about a (mathematical) scalar field that makes is particularly suited to represent a gravitational potential as opposed to, say, a Higgs field. We often use the very same mathematical objects to represent very different things, and what they represent is a matter not of their mathematical structure *per se* but how we are using the mathematics to represent the world. *This* orientation represents the passage of time simply because we have decided to so use it, and another (mathematically identical) orientation can be used to represent something quite distinct.

Mathematical physics is a field that uses mathematical objects as *representations* of physical states of affairs. And as with all representations, the content of the representations depends on certain representational *conventions*: particular mathematical objects are stipulated to represent particular physical items. One cannot recover these conventions by simply examining the representations themselves: one must have additional knowledge of the conventions. If one overlooks this basic fact, all sorts of confusions will arise.

To begin with a simple example: the physicist uses purely mathematical objects such as manifolds with Lorentzian metrics, to represent physical spatio-temporal structure. This requires a choice of conventions, such as whether Minkowski space-time is represented by a metric whose signature is $(+,-,-,-)$ or $(-,+,+,+)$, i.e. whether timelike directions are to be represented by vectors whose norms are positive or negative. This is clearly a matter of pure conventional choice. Of course, if one *knows* already that it is Minkowski space-time that is being represented, one can tell which convention has been chosen by examining the mathematical structure, but the convention is still in play. This is sometimes forgotten when dealing with two-dimensional space-times, where the metric will be of signature $(+,-)$. It is tempting to conclude that in such a space-time the spacelike and timelike directions have become somehow inherently indistinguishable because one cannot read off the convention from the mathematical object. But this is not to be expected: if a survey result is reported by using columns headed 'P' and 'Q' rather than 'Yes' and 'No', that does not mean that the *answers* are somehow indistinguishable: one simply has to know the convention being used in reporting the results. Similarly, when using a two-dimensional mathematical object to represent a two-dimensional relativistic space-time, one simply has to know what representational conventions are being used to distinguish the spacelike from the timelike directions.

Matters can get even more confusing when one uses *space-time diagrams* to represent relativistic spatial-temporal structure. The paper on which the diagram is drawn is typically an isotropic material. Cut into rectangles, the edges of the paper pick out directions as horizontal and vertical. Clearly, if some directions of lines on the paper are to represent timelike directions in space-time, others spacelike directions, and yet others null directions, conventions must be adopted. Knowing those conventions, it is not hard to interpret the diagrams. But there is no reason to suppose—as is sometimes done—that one ought to be able to recover those conventions from a close inspection of the diagrams themselves. For example, Brad Skow, in the course of asking 'What Makes Time Different from Space?' (forthcoming), presents us with a space-time diagram stating 'Let's take a look at a particular world with a two-dimensional space-time and see if we agree that the laws distinguish space from time even though the geometry of spacetime does not' (section 6). But of course what we are presented with is not a particular world, but a space-time *diagram*. The geometry of the *diagram*, which is essentially Euclidean, surely does not tell us which directions on *it* represent space-like directions and which time-like. But the requisite information is not to be sought in further details of the diagram, it is to be sought in the representational conventions. The geometry of the *space-time represented* does distinguish the space-like and time like-directions, even apart from the material contents of the space-time.

But there seems to be a deeper problem. The proper mathematical representation of the direction of the passage of time (i.e. the direction from past to future) may be an orientation on the mathematical manifold representing the space-time. This orientation does provide a temporal anisotropy, but, one might say, there is nothing about it that suggests movement or change or flow at all. It is completely static.

The obvious answer to this worry is that orientations on mathematical manifolds are static and unchanging exactly because all mathematical objects are, in their own nature, 'static' in the sense of being outside of time and unchanging. And because of this, one will always have the feeling that mathematical objects are, in their own nature, not adequate to represent the intrinsic nature of time. This is presumably the reason that we commonly talk of attempts to 'spatialize' time when we do mathematical physics, but never hear about attempts to 'temporalize' space. This arises, I think, simply from the choice of mathematical objects as representations. (Of course, mathematical objects are no more really spatial than they are temporal, but we are so accustomed to think about mathematical objects by contemplating *diagrams* that we can't shake the notion that they have something like spatial structure.) We could, of course, choose to use temporally structured *physical* objects to represent purely spatial structures, and so 'temporalize' space. It might be therapeutic to consider how this could be done.

Suppose our physics, instead of using mathematical objects to represent physical reality, used instead some intrinsically time-directed representational medium. For example, suppose one used musical tunes to represent things. A collapsing star would be indicated by a falling tone (under the convention that the lower the pitch, the smaller the radius of the star), while a supernova would be represented by a suddenly rising one. Such a scheme would have some obvious advantages when used to represent processes: since the representations themselves are time directed, the representations of any sequence and its time reverse would be obviously different. Using a spatial chart to track, e.g. one's weight through time needs something extra to indicate time direction: one might know that one axis represents weight and the other time, but still not know if the chart represents weight gain or weight loss. (The problem is worse if the charts are kept on panes of glass, with no indication of which side is 'front': then a weight-loss chart can exactly coincide with a weight-gain chart if the time direction arrow is omitted.) If we used tunes, then the two charts could never be confused with one another.

But, of course, this representational scheme would have corresponding difficulties when it comes to representing purely spatial facts. Suppose, for example, one wanted to record the elevation profile of the Rocky Mountain chain. Looking at the chain, one might, as it were, sing the profile as a tune of rising and falling pitch. But the nature of the tune—the way it sounds—would differ greatly depending on which direction one scans the chain: left to right or right to left, or whether one sees the chain from the east or the west. (It helps to imagine that this culture has developed no musical notation, so a convention for 'reading notes' does not exist, and to imagine that written records are read up to down. We are so accustomed to spatialized representations of temporal sequences that we have internalized the conventions which relate spatial characteristics of the representations to temporal aspects of the thing represented.) So suppose some early Lewis and Clark come back from an expedition and are asked to describe the mountain chain they have seen. Lewis whistles one tune, and Clark the same tune in retrograde motion.[13] These tunes might well, to the untrained ear, sound as if they have nothing at all in common. But Lewis and Clark would insist that the profile of the mountain chain is that non-temporal object that is equally well represented by either of these two quite different-sounding tunes.

One might say that there is something *infelicitous* about this way of representing spatial facts: representations that would strike one as radically different are supposed to have the same content, and it might even be quite difficult to tell when two representations represent the same spatial facts (again, suppose they have no musical notation). Furthermore, consider how someone who felt perfectly at home with tunes might raise philosophical problems about the idea of space: 'You tell me that spatial facts are the sort of facts that can with equal fidelity be represented by either of a pair of completely different-sounding tunes. But I hear nothing at all in common between the tunes: they sound completely different. The idea of something they have in common is abstract and unintuitive, unlike the tunes themselves, which I can hear. So I just don't understand what this "thing in common" is supposed to be, or how there could be something that is indifferently represented by two such distinct representations.' The problem, of course, is that we are being asked to ignore the temporal-directedness of a representation whose temporal-directedness is quite salient, so we feel that the representation does not really fit to its object, even though the representational scheme can be described precisely.

Similarly, I think, we can have the lingering feeling that mathematical objects *per se* are not fit to represent the passage of time, even when the

---

[13] It is notable that the one tune would be the other in retrograde motion: we can easily 'hear' that one tune is the *inversion* of another, but it is much more difficult to recognize a tune played backwards.

convention relating the passage to the mathematics has been completely fixed. We are given an orientation, and told which set of light-cones are the future light-cones and which the past, so we can now distinguish representations of Earth-to-Mars asteroids from Mars-to-Earth asteroids, and distinguish representations of weight losses from representations of weight gains, and can distinguish representations of normally functioning humans from bizarre anti-metabolizing, anti-thermodynamic things. Perhaps a mere orientation seems somehow inadequate to capture the difference between these, but the apparent inadequacy must be an illusion. Of course, we could not use the pure mathematics to get someone who did not understand what the passage of time is to understand it, any more than we could use tunes to get someone who did not understand what space is to understand it. But everyone is perfectly familiar with the passage of time, and we would not really know what to make of someone who claimed not to comprehend the idea that some events lie to the future and some are in the past, or that there is a difference between losing and gaining weight, or traveling north as opposed to south. If such 'time blindness' really could exist, we probably could not remedy it by trying to teach the person physics. But none of that gives us any reason to question the passage of time.

In sum then, it is a central aspect of our basic picture of the world that time passes, and that in virtue of that passage things change. And there are no good logical or scientific or philosophical arguments that cast doubt on the passing of time, and there are no impediments to representing, in our present physical theories, that time passes. I draw what ought to be a most uninteresting conclusion, but one that has somehow managed to become philosophically bold: time does pass. Its passage is not an 'illusion' or 'merely the product of our viewpoint' or 'an appearance due to our special mode of perception'. Its passage is not a myth. The passing of time may be correlated with, but does not consist in, the positive gradient of entropy in the universe. It is the foundation of our asymmetrical treatment of the initial and final states of the universe. And it is not to be reduced to, or analyzed in terms of, anything else.

5
Causation, Counterfactuals, and the Third Factor

In his classic paper 'Causation' (1973b), David Lewis urged that regularity accounts of causation, having failed over a long period of time to resolve the counter-examples and problematic cases with which they were faced, be abandoned in favor of an alternative strategy: the analysis of causation in terms of counterfactuals. More than a quarter of a century later, the time has come to evaluate the health and prospects of Lewis's program in its turn.

No one would now deny that there is some deep conceptual connection between causation and counterfactuals. As Lewis points out, Hume himself (incorrectly) paraphrased one of his analyses of causation using a counterfactual locution:

… or in other words, *where if the first had not been, the second had never existed.*

In honor of Hume, let us call the counterfactual 'If C had not occurred, E would not have occurred', when both C and E did actually occur, the *Hume counterfactual*. When we think we know a cause of some event, we typically assent to the corresponding Hume counterfactual. Furthermore, our interest in causes often has a practical aspect: we want to know the causes of events so that we can either prevent or foster similar sorts of events at other times. Causal claims are therefore deeply implicated with the sorts of future subjunctives used in practical deliberation: if we should do X (which we might or might not, for all we now know) then the result would be Y. The future subjunctive is a close cousin to the counterfactual conditional, since accepting a future subjunctive commits one to accepting the corresponding Hume counterfactual in the event that the antecedent does not come about.

But Hume's dictum hardly survives a few minutes' contemplation before counter-examples crowd into view. The sort of counterfactual dependency Hume cites is not necessary for causation: perhaps the effect still would have occurred despite the absence of the cause since *another* cause would have stepped in to bring it about. The dependency is also not uncontroversially sufficient for causation. If John Kennedy had not been assassinated on 22 November 1963, he would have still been president in December 1963. But surely too, if Kennedy had still been president in December 1963, he would not have been assassinated in November of that year: when asked to consider the latter counterfactual we do not imagine Kennedy killed in November and resurrected in December. So the counterfactual dependencies go both ways and the causal arrow only one.

These sorts of problems have been thoroughly chewed over in the literature, and no detailed review of the strategies used to respond to them is needed here. The problem of back-up (potential) causes can perhaps be solved by using counterfactual dependency to define direct causation and then taking the ancestral to define causation, or by making the effect *qua* effect very fragile in some way and arguing that the back-up could not have produced it at exactly the right time or right place or with all of the right properties, or by a matching strategy, in which the cause which is operative when the back-up is present is identified by matching it to an otherwise identical case where the back-up is absent (and the Hume counterfactual holds). Attempts to fix up the counterfactual analysis have become ever more subtle, and complicated, and convoluted. I have neither the space not the desire to address them all individually.

I do not want to quarrel with these sorts of solutions in detail because I want to argue that the attempt to analyze causation in terms of counterfactuals of this sort is wrong-headed in a way that no amount of fine-tuning can fix. Causation is not to be analyzed in terms of counterfactual dependency at all, no matter how many equants and epicycles are appended to the original rough draft.[1]

If causation is not to be analyzed in terms of counterfactual dependency, how are we to explain the systematic connections between judgements about causes and judgements about counterfactuals? Such connections may be secured analytically if causation can be defined in terms of counterfactuals, but I am denying this can be done. Connections would also obtain if counterfactuals could be analytically reduced to causal claims, but that is an even less appetizing project than the one we are rejecting. The only other possibility

---

[1] Ned Hall has suggested that there are really two concepts of causation, one of which ('dependence') amounts to nothing more than the Hume counterfactual, and the other ('production') cannot be so analyzed (cf. 'Two Concepts of Causation', chapter 9 of Collins, Hall, and Paul 2004). If one wishes, this paper can be read as an account of production, although I suspect that at least many cases of dependence can be covered using these methods.

is that a third factor is involved, some *component* of the truth conditions of counterfactual claims that is also a component of the truth conditions for causal claims. This third factor would provide the analog of a 'common cause' explanation for the systematic connections between causal claims and counterfactuals: neither underpins the other but the third factor underpins them both.

The prospects for a 'third factor' explanation obviously depend on the identification of the missing factor. I think it can be identified: what links causation and counterfactuals by figuring in the truth conditions for both is natural law. Laws play one role in determining which counterfactuals are true, and another role in securing causal connections. The 'necessary connexion' that Hume sought at the heart of causation is nomic necessity.

1. KNOWLEDGE OF CAUSATION WITHOUT KNOWLEDGE OF ANY HUME COUNTERFACTUAL

Let us consider what one must know, in at least one case, in order to know what caused an event. Suppose you know that the laws that govern a world are the laws of Newtonian mechanics. And suppose you also know that forces in this world are all extremely short range: forces only exist between particles that come within an angstrom of each other. And suppose particle P is at rest (in an inertial frame) at t_0 and moving at t_1, and that in the period between t_0 and t_1 only one particle, particle Q, came within an angstrom of P. Then, I claim, we know with complete certainty what caused P to start moving: it was the collision with Q.

Thus: given that we know the laws and we know some very circumscribed set of particular facts, we know the cause of P's motion. But do we know what would have happened if Q had not collided with P (i.e. if Q had not approached within one angstrom of P)? We do not. Suppose, for example, that the vicinity of our particles is chock-a-block full of monitoring equipment, which tracks the exact trajectory of Q, and jammed with particle-launching devices loaded with particles just like Q and designed to launch these particles so as to collide with P just as Q did if Q should deviate in any way from its path. There are (for all I have told you) innumerable such devices, monitoring the path at arbitrarily close intervals, all prepared to step in should Q fail to collide with P. I hereby warn you of the presence of these contraptions, unspecified in number and construction, and ask you now whether we know what would have happened if Q had not collided with P.

We do not have enough information to evaluate this counterfactual, both because the *way* Q fails to collide has not been specified and because the exact construction and disposition of the monitoring devices has not been indicated. Perhaps for many sorts of deviation, some other particle, just like Q, would have collided with P at just the same time and in just the same way. Perhaps this is even true for *all* possible deviations, or all the sorts of deviation we would consider relevant. So perhaps we could be convinced that P would have moved at exactly the same time and in just the same way even if Q had not collided with it. Still, *none of this has anything at all to do with the fact that the collision with Q is what caused P to move.* The existence of the monitoring devices and potential back-up particles is simply *irrelevant* to the claim that the collision with Q was the cause. In fact, once we know the laws we don't even *care* what would have happened if Q had not collided with P: Perhaps P would not have moved or perhaps it would have because something else collided with it. The information we have (namely the laws of nature in this world) allows us to identify the cause without knowing that *any* Hume counterfactual is true.

The counterfactual analyst can respond in several ways. One would be to insist that we know the causes in this case because we know that *some* Hume counterfactual is true, although (due to the lack of information) we don't know which one. No matter how many back-up systems there are, there must be some point on the trajectory of Q such that had it miraculously swerved at that point, P would not have moved since none of the back-ups would have had a chance to fire. But since I have not specified the number or location of the back-ups, how does one know this is true? What if there is an infinite sequence of back-ups, progressively faster and faster reacting, monitoring at ever closer and closer intervals?

Lewis addresses this sort of criticism in appendix E to 'Causation', claiming that we need not pay much heed to our intuitions about such recondite circumstances. Discussing exactly this sort of problem (as raised by William Goosens) and a case involving action at a distance, he responds:

I do not worry about either of these far-fetched cases. They both go against what we take to be the ways of this world; they violate the presuppositions of our habits of thought; it would be no surprise if our common-sense judgments about them went astray—spoils to the victor! (1986a, p. 203)

Lewis's strategy here deserves some comment. The rules of the game in this sort of analytic project are relatively clear: any proposed analysis is tested against particular cases, usually imaginary, for which we have strong intuitions. The accuracy with which the judgements of the analysis match the deliverances of intuition then constitutes a measure of the adequacy of the analysis. Unfortunately, it is often the case that the question of *how* the intuitions are arrived at is left to the side: getting the analysis to deliver up the right results, by hook or by crook, is all that matters, and this, in turn, encourages ever more baroque constructions. But if we care about intuitions at all, we ought to care about the underlying mechanism that generates them, and in this case it seems very plausible that a simple *argument* accounts for our judgement, *an argument entirely unaffected by the existence of back-up devices, no matter how numerous*. The case does differ from those of our actual experience, and in such a way that the envisioned situation will surely never occur. But the *way* that the example differs from more familiar ones *makes no difference whatsoever to the reasoning that allows us to identify the cause*: so long as none of the back-up devices *fires*, their number and complexity are perfectly irrelevant. We have such strong intuitions about the case even though it is far-fetched because the respect in which it is far-fetched makes no difference to the method by which the cause is identified. The example shows that causes are not identified via the Hume counterfactual.

What one would like to say, of course, is that the following counterfactual it true: If *Q had not collided with P and none of the back-up devices had fired*, then *P* would not have moved. And this is correct. So we want to hold fixed, in evaluating the counterfactual, *the non-firing of potential alternative causes*. But in order to know what to hold fixed, we already have to know quite a lot about the (actual and counterfactual) causal structure, since we have to identify the possible-but-not-actual-alternative-causes to hold fixed. But now the project looks pretty hopeless as a way of using counterfactuals to get a handle on causation: we already have to bring in causal judgements when determining which counterfactual to consider.

So if we do not know that the collision with Q caused P to move by knowing what would have happened if Q had not collided with P, how do we know it? The argument is quite simple and straightforward, but is worth laying out. Since the laws of this world are by hypothesis those of Newtonian physics, we know that particle P, which is at rest at t_0, will remain at rest unless some force is put on it. And since the forces in this world are all shortrange, we know that no force will be put on P unless some particle approaches within one angstrom of it. And since P does start to move, we know that some particle did approach within one angstrom of it, which caused it to start moving. And since *Q* was the *only* particle to approach closely enough, we know that it was the collision with *Q* that caused *P* to move. End of story.

There are some counterfactuals implicit in this reasoning. If *nothing* had put a force on *P* it would not have accelerated. That, of course, follows from Newton's First Law of Motion. And if we knew that if *Q* had not put a force on *P nothing else would have*, then we would have the Hume counterfactual. But we do not know this. So as it stands we can only infer: if *Q* had not collided with *P* either *P* would not have started moving or something else would have collided with it. This is not the Hume counterfactual, hence the headaches for the counterfactual analysis.

Various sorts of sophistication of the original simple counterfactual analysis might succeed in getting the case right, once enough detail is given. For example, if *Q* had not collided with *P*, but some other particle had instead, then it does seem plausible that *P* would not have moved off in quite the same way as it did. This seems plausible even though we cannot begin to specify *how P*'s motion would have differed. It also seems plausible that there is *some* moment in *Q*'s trajectory late enough that, had *Q* miraculously disappeared, it would be too late for a back-up to fire—even though we have no clue about *how late* that point must be. So maybe these sorts of analysis can get the right result. But the curious thing is that even if they do get the right result, *it is obviously not on the basis of that analysis that we judge Q to be the cause, rather it is because we already judge Q to be the cause that we think that, with enough detail, we would get the right result*. This is obvious since our judgement that *Q* is the cause is so secure *even before the details are filled in*. And the advantage of the little argument rehearsed above is that it explains how we know that *Q* was the cause, given the laws of motion and the fact that only *Q* collided with *P*, without having to know any more details about the situation. So even if it does turn out that had *Q* not collided, *P* would have at least moved differently, that is irrelevant (given the laws) to identifying the cause.

Since it is facts about the laws that help us identify the cause in this case, and since laws are obviously deeply implicated in the evaluation of counterfactuals, I suggest that we stop trying to analyze causation directly in terms of counterfactuals and consider anew how laws play a role in determining causes. At one end of the spectrum we have seen how the laws of Newtonian mechanics can enable one to identify the cause in the example discussed above, even if one does not know any Hume counterfactual. Let us now look at an example from the other end of the spectrum.

2. KNOWLEDGE OF COUNTERFACTUALS WITHOUT KNOWLEDGE OF CAUSES

If an analysis of causation in terms of counterfactuals were correct, then, unless the analysis itself contains some vague concepts, fixing determinate truth values for all counterfactuals should fix the truth values of all causal claims. Of course, in many circumstances counterfactuals themselves may not have classical truth values: the antecedent or consequent of the conditional may be too vague. But supposing that the truth values of all relevant counterfactuals are sharp, then one would suppose that an analysis of causation in terms of counterfactuals should render the causes sharp. But we can imagine situations in which all relevant counterfactuals are determinate but causal claims are not, because it is unclear exactly what the laws are. The example is rather artificial, but illustrative.

Recall the rules of John Conway's game of Life. Life is played on a square grid, using discrete moments of time. At any moment, each square in the grid is either empty or occupied. At any given moment of time, whether a square is occupied or not depends on how that square and the eight immediately adjacent squares were occupied at the previous moment. For example, if four or more of the adjacent squares are occupied at one instant then the given square will be empty at the next instant, and if one or fewer of the adjacent squares are occupied then at the next moment the square will be empty. Conway's rules cover all possibilities, so the state of the grid evolves deterministically through time.

Now imagine a world in which space and time are discrete, and in which space consists in a rectangular grid of points that persist through time. And imagine that there is only one sort of matter in this world, so that every discrete part of space is, at any moment, either empty or occupied by the matter. And imagine that the patterns of occupation evolve in definite deterministic patterns, similar to those in Conway's game, but somewhat less orderly. As in Conway's game, one can predict with certainty whether a given point will be occupied or not at an instant if one knows only the previous pattern of occupation of that point and the eight adjacent points. But unlike Conway's game, the rules of evolution cannot be distilled into a few simple rules.

In principle, the rules of evolution must specify for each of the 512 possible patterns of occupation of a three-by-three grid whether the central square is

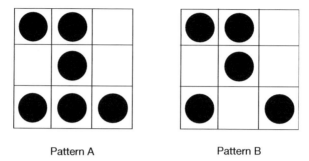

Figure 9. Two patterns that yield the same outcome

or is not occupied at the next instant.[2] Imagine picking the rules in a random way: for each of the 512 possible input patterns, flip a coin. If the coin lands heads, the central square will be occupied the next instant, if tails it will be empty. One now has a complete set of rules that will determine how patterns evolve through time. Now imagine a physical world of infinite extent and with a completely variegated state of occupation whose evolution in time always conforms to these rules.

(It is a key point that we are now imagining a world whose evolution everywhere *conforms to* these rules, but we are reserving judgement about whether we ought to say that the evolution is *generated by* these rules.)

If we have chosen the rules randomly, then it is overwhelmingly likely that there will be patterns of occupation which differ only in a single location but which both yield the same result. For example, suppose it turns out that both patterns *A* and *B* in Figure 9 are always succeeded by the central square being occupied.

The question I want to ask it this: in such a world, is the bit of matter in the bottom central location of pattern *A* a *cause* (or an essential part of a cause) of the central square being occupied the next instant or not?

In the course of this enquiry, I take a few points to be uncontroversial. First, that the sort of physical world I have described at least *could be* a world that has laws of the evolution of the distribution of matter. On anything like the Mill–Ramsey–Lewis account of laws, I take it that if the world is extensive and variegated enough, then there *must be* laws: after all, of all possible distributions of matter throughout space and time, only a set of

---

[2] If one imposes some natural symmetry constraints, such as 90-degree rotational symmetry, then there will be fewer than 512 distinct cases to deal with.

measure zero will conform to these rules through all time. It is therefore extremely informative to know that these rules are always obeyed. I further assume that having granted that the world is law governed, we grant that the truth values of all counterfactuals concerning the evolution of distributions of matter are determined. Since the rules are deterministic, we can use them to determine how any distribution of matter would have evolved, even if that distribution never actually occurs in the world we are imagining. So at the fundamental physical level, there is no dispute about counterfactuals: given any complete pattern at a time, we know how it would have evolved had it been different in some precisely specified way.

My central claim is this: even though there is no dispute, at the fundamental level, about counterfactuals, there can still be a dispute about causation, and further, this dispute about causation arises as a consequence of a parallel dispute about laws. The issue here is not how the case ought to be judged, or what our intuitions are about the case, or whether a particular theory gets the intuitions right. The issue is rather that unproblematic knowledge of all counterfactuals in this case does not seem to settle the question of causal structure, and furthermore unproblematic knowledge of counterfactuals does not settle what the laws are in this same case. This again suggests that it is the laws, not the counterfactuals *per se*, which underwrite causal claims.

There are two lines of argument that can be addressed to the question whether the bit of matter in the bottom central location of pattern *A* is a cause of the succeeding occupation or not. The first line makes straightforward appeal to the Hume counterfactual, and denies the matter in that location is a cause. It is beyond dispute in this case that the Hume counterfactual is false: even if the bottom central location had not been occupied (i.e. even if what was pattern *A* had been instead pattern *B*), still the central square in the succeeding instant would have been occupied. In this sense, it is beyond dispute that the presence or absence of matter in that location makes no difference to how the distribution of matter will evolve. Furthermore (this argument continues) although sometimes the Hume counterfactual can fail to be true for genuine causes and effects, that is only because causation is the ancestral of *direct* causation, and in direct causation the Hume counterfactual is always true. But since space and time are discrete here, and there are no intervening moments of time or intervening places in space, if there is any causation here it must be direct causation. There is simply no place for there to be a chain of direct causes between pattern *A* and the succeeding occupation of the central square. So since there is no indirect causation (no room for it) and there is no

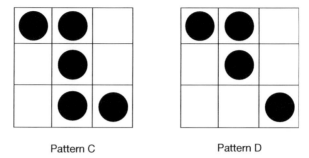

Pattern C   Pattern D

Figure 10. Two patterns that yield different outcomes

direct causation (the Hume counterfactual is false) there is no causation at all linking the matter in the bottom central location to the matter occupying the central square in the next instant. Thus the case against causation.

Now the case for causation. In arguing in favor of the matter in the bottom center being a cause, one would first point out that in general the laws of evolution in this world seem to link complete three-by-three patterns of occupation to the succeeding state of the central square. If the state of the central bottom location were *never* relevant to the evolution, then there would be little dispute about its causal efficaciousness. But suppose that patterns C and D in Figure 10 are such that C is always followed by an occupied central square and D by an empty one.

In this case, the Hume counterfactual holds, and the matter in the bottom central square of pattern C would be among the causes of the occupation of the central square in the next instant. Now the idea is that the complete set of 512 transition rules is so miscellaneous that there is no significantly *more* compact way to convey it than by specifying each of the 512 rules separately. It does not, for example, make things significantly simpler to try to bundle patterns A and B together into *one* rule (which refers to only the pattern of occupation of the eight locations other than the bottom central one) since one would have to indicate, either explicitly or implicitly, that in this case the state of the bottom central location need not be specified, although in other cases it must be.

So the advocate of causation argues as follows. The overall structure of the evolution in this world suggests that the laws of nature connect complete three-by-three patterns to the succeeding state of the central square. (After all, that is how *we* generated the rules: by flipping a coin anew for each of the 512 patterns.) Since there is no interesting generic pattern to the rules, the most *uniform* thing to do is to regard the fundamental laws of this world as all rules that take complete three-by-three patterns as input and give either an occupied or empty central square as output. From this point of view, it is just a *coincidence* that pattern *A* and pattern *B*, which differ only in the state of the bottom central square, both lead to occupied central squares. (After all, if we generate the rules by flipping a coin, it *is* just a coincidence that they do.) As far as the laws of nature are concerned, pattern *A* and pattern *B* have *nothing in common at all*, they are, as it were, each atomic and distinct. From this point of view, then, the state of the central bottom square is of vital importance, for it determines whether the transition falls under the law governing pattern *A* or the law governing pattern *B*, where these *laws* are regarded as fundamentally distinct. If an *A* pattern gives rise to an occupied central square, then while it is true that the square would still have been occupied had the bottom central location been empty, that would have been a case of an *alternative and distinct cause of the same effect*. The *B* pattern is a sort of back-up mechanism, on this view, which would produce the same effect as the *A* pattern had the *A* pattern been different in a specified way. But whether the transition from the *B* pattern counts as an *alternative mechanism* or as an instance of the *same* mechanism as the transition from the *A* pattern depends on what the laws are that govern the situation. Both transitions could be instances of the operation of the same law, or instances of two distinct laws, and our causal judgements depend on which we take to be the case.

We saw above that one route to save the counterfactual analysis was to add something to the antecedent: that all potential back-up causes be held fixed as non-firing if in fact they did not fire. We saw that this spells trouble for trying to use the counterfactuals to analyze causation. Now we are seeing that what counts as an alternative or back-up cause may depend on the nature of the laws governing the situation, so it is not puzzling that analyses that leave the laws aside will run into trouble.

I do not wish to adjudicate between the arguments pro and con the causal effectiveness of the bottom central location in this case, I rather simply want to acknowledge that there can be a dispute, a *reasonable* dispute, in a case like this where the counterfactuals are all beyond question. I do think, as a matter of fact, that the pro causation argument would eventually win out, if no interesting generic meta-rules could be found which simplified the presentation of the 512 different transitions. But even if one thinks that the con side would win, it is hard to deny that each side *has a point*, and that the point ultimately turns on how one is to conceive of the laws of nature in such a world.

We can easily imagine the 512 different transitions being such that they can be encapsulated by very simple rules, such as the Conway rules mentioned above. And we can imagine intermediate cases, where instead of giving all 512 transitions separately, one can give a few, very complicated, generic rules. And we can imagine the case discussed above, where no interesting simple meta-rules exist. There is a slippery slope between these cases. I take it that on Lewis's account of laws, there will be points along this slope where it will be indeterminate what the laws are: points where there will be reasonable disagreement about whether the fewer complicated generic rules are *simpler overall* than the 512 separate transition rules. And where the laws are in dispute, it may turn out (as it does above) that the *causes* are in dispute, all while the truth values of the counterfactuals remain unquestioned.

So our pair of examples displays a nice symmetry. In the Newtonian case, we showed that we can know the causes without knowing any Hume counterfactuals if only we know the laws. And in our modified game of Life, we have shown that we can know all the counterfactuals without being sure about the causes if there is a dispute about the laws. Between the two of them, they make a strong case for the idea that it is the laws rather than the counterfactuals which determine the causes.

3. THE ROLE OF LAWS

If we are to explain the ubiquitous connections between causal claims and counterfactuals by appeal to the role of laws in the truth conditions of each, we must at least sketch how those truth conditions go. In the case of counterfactuals the basic story is extremely obvious and requires little comment. If I postulate that a world is governed by the laws of Newtonian mechanics, or the laws of the game of Life, then certain conditionals are easy to evaluate: namely those that postulate a particular complete physical state at a given time as the antecedent and some other particular or generic physical state at a later time as the consequent. Given the state specified in the antecedent, the laws then generate all later states, and the sort of state specified by the consequent either occurs or does not. That is, one uses models of the laws as possible worlds and then employs the usual possible-worlds semantics for the truth conditions of the conditional.

If the antecedent is not stated directly in the physical vocabulary, or is vague or generic, or is only a partial specification of the state of the world, then various sorts of filling in must be employed to yield a set of states which satisfy the antecedents, which then generate a set of models, and again things proceed in the usual way. There is much to be amplified in this sketch, and emendations needed for probabilistic laws, etc., but the role of laws in the whole process is evident enough. Let us omit any more detail about counterfactuals, then, and turn to the much more interesting topic of causation.[3]

I do not think that there is any *uniform* way that laws enter into the truth conditions for causal claims, as there is a fairly uniform way they enter into the truth conditions for counterfactuals. Rather, I think that laws of a very particular form, wonderfully illustrated by Newton's laws of motion, support a certain method of evaluating causal claims while a more generic, and somewhat less intuitive use must be found for other sorts of laws. As we will see, the laws of the Life world lack all of the interesting characteristics of Newton's laws, so we will again use them for illustrative purposes.

Let's start, then, by returning to our colliding particles. Why are we so confident in identifying the collision with particle Q as the cause of particle P being set in motion? Recall the exact form of Newton's laws, the form that makes them so useful as aids for tracking causes in a case like this. The First Law, the so-called *Law of Inertia*, states that a body at rest will remain at rest and a body in motion will continue in motion at a uniform speed in a straight line, unless some force is put on it. The first law specifies *inertial motion*, i.e. how the state of motion of an object will progress if nothing acts on it. The second law then specifies how the state of motion of an object will *change* if a force is put on it: it will change in the direction of the force, and proportionally to the force, and inversely proportionally to the mass of the object. Note that the Second Law is parasitic on the First: the First specifies what is to count as a state of motion (uniform motion in a straight line), and the Second how, and in what circumstances, the state changes.

The structure of Newton's laws is particularly suited to identifying causes. There is a sense, I think, in which the continuation of inertial motion in a Newtonian universe *is not caused*. If a body is at rest at one time, and nothing acts on it (i.e. no force acts on it), then it sounds odd to ask what causes it to remain at rest. It sounds odd to say that the body's own inertial mass causes it to remain at rest, since there is no force which the mass is resisting, and the inertial mass is just a measure of a body's resistance to force. And it sounds

---

[3] More detail can be found in 'A Modest Proposal Concerning Laws, Counterfactuals, and Explanations', Chapter 1, this volume.

odd to say that the Law of Inertia itself causes the body to remain at rest, since it seems similar to a category mistake to ascribe causality to the very laws. Of course, the body remains at rest because no force acts, and because the inertial state of motion of a body at rest is to remain at rest, but it certainly sounds odd to cite the absence of forces as a cause of remaining at rest.

Or at least, if there is any cause of a body at rest subject to no force remaining at rest in a Newtonian universe it is a sort of *second-class* cause: the first-class Newtonian causes are forces (or the sources of forces), and what they cause, the first-class form of a Newtonian effect, is a change or deviation from an inertial state of motion. There is no doubt that for Newton, once the First Law is in place, that one can ask what causes the Earth to orbit the Sun (rather than travel at constant speed in a straight line), and that the cause in this case is the gravitational force produced on the Earth by the Sun. It is this sort of conceptual structure that allows us to so readily identify the cause of the motion in particle P above: since the inertial state of P is for it to remain at rest, its change into a state of motion requires a cause, i.e. a force, and the only force is provided by Q. Without the law of inertia, it would not be clear that the sudden onset of motion in P required any cause at all: perhaps particles sometimes just spontaneously start moving. Or perhaps the inertial state of motion of a particle could be a jerky motion: the particle moves at a given velocity for a while, then stops for a while, then resumes at the same velocity, so the onset of motion in P is just the continuation of its natural state of motion.

Let us denominate laws *quasi-Newtonian* if they have this form: there are, on the one hand, *inertial* laws which describe how some entities behave when nothing acts on them, and then there are laws of *deviation* which specify in what conditions, and in what ways, the behavior will deviate from the inertial behavior. When one conceives of a situation as governed by quasi-Newtonian laws, then typically the primary notion of an effect will be the deviation of the behavior of an object from its inertial behavior, and the primary notion of a cause will be whatever sort of thing is mentioned in the laws of deviation.

Laws, of course, need not be quasi-Newtonian. At a fundamental level, the laws of the game of Life are not. In that game, if a cell is occupied at one moment then it has no 'natural' or 'innate' tendency to remain occupied—or to become unoccupied—at the next moment. The patterns evolve in an orderly way, but the behavior of the overall pattern cannot be analyzed into, on the one hand, inertial states of the parts and, on the other, interactions which cause deviations from those states. There is no division of the rules of evolution into the inertial rules and the rules for change of inertial state.

That does not, of course, mean that the Life world is without causes. There are causes and effects, but the way we conceive of them, at the fundamental level, is quite different from the way we think of causes in a quasi-Newtonian setting. If a particular cell in the Life world (or, more obviously, the modified Life world) is occupied at a moment, then the cause of that is the complete three-by-three pattern of occupation centered on that cell the instant before, and similarly if a cell is empty. And the cause of that three-by-three pattern a moment before is the complete five-by-five pattern of occupation centered on that cell two moments before, and so on. As we go further back in time, the reach of the ancestral causes grows outward like a ziggurat, the Life world analog of the past light-cone of an event in a Relativistic universe. At a fundamental level, the only proper causes in the Life world are these complete patterns at a time which—in conjunction with the laws—generate the successive patterns that culminate in the given cell being occupied or empty at the given moment. (In the context of the discussion I am assuming that we agree that only a full three-by-three pattern is a cause of the succeeding state of the central square.)

This notion of causation will be available in any world governed by deterministic laws, whether quasi-Newtonian or not, and so the philosopher hankering after the most widely applicable concept of causation will likely be drawn to it. It is obviously an instance of the INUS concept of causation, for example. But I think that we only fall back on this notion of causation when circumstances demand it: our natural desire is to think of the world in quasi-Newtonian terms, in terms of inertial behavior (or 'natural' behavior) and deviations from inertial behavior: in terms of, to use a concept from mathematical physics, perturbation theory. Largely this is because it is much easier to think in these terms, to make approximate predictions on the basis of scanty data, and so on. And often circumstances allow us to think in quasi-Newtonian terms even when the underlying laws are not quasi-Newtonian, or to think in macro level quasi-Newtonian terms quite different from the laws that obtain at the fundamental level. Indeed, the search for quasi-Newtonian laws does much to explain the aims of the special sciences.

4. CAUSATION AND MACROTAXONOMY IN THE SPECIAL SCIENCES

I am a realist about laws: I think that there are laws, and that their existence is not a function of any human practices. I am also a primitivist about laws: I do not think that what laws there are is determined by any other, distinctly specifiable set of facts, and that in particular it is not determined by the total physical state of the universe. And I am a physicalist about laws: the only objective primitive laws I believe in are the laws of physics. Speaking picturesquely, all God did was to fix the physical laws and the initial physical state of the universe, and the rest of the state of the universe has evolved (either deterministically or stochastically) from that. Once the total physical state of the universe and the laws of physics are fixed, every other fact, such as may be, supervenes. In particular, having set the laws of physics and the physical state, God did not add, and could not have added, any further laws of chemistry or biology or psychology or economics.

We do not, however, understand the vast majority of what we do understand by reflecting on the laws of physics. For example, there is much that I understand about how the computer I am now using works, and precious little of that derives from detailed knowledge of the physics of the machine. Rather, I understand its operation by thinking about it in terms of some *lawlike generalizations*, generalizations which resemble laws at least insofar as being regarded as supporting counterfactual claims and being confirmed by positive instances.

In this sense, it is a lawlike generalization about the computer that when it is turned on and the word processing program is running and there is a document open, pressing a key on the keyboard will be followed by the corresponding letter appearing at the point where the cursor is, and the cursor will move over one space to the right (unless it is the end of a line, in which case it moves all the way to the left on the next line, unless it is at the bottom of the window, etc., etc., etc.). This generalization, which could be made more precise and extensive at the cost of much tedium, is taken to support counterfactuals: if I had hit the letter 'z' on the keyboard instead of 's' just before the last colon, the word which would have appeared would have been 'counterfactualz'.

No doubt it is by means of such generalizations that we understand how to use the computer, predict how it will function, explain it to others, etc. And no doubt this generalization, albeit lawlike in certain respects, is not in any metaphysically interesting sense a law. If I should hit a key and the corresponding letter did not appear, then it is not that any *law* would be broken: rather *the computer* would be broken (or misprogrammed, or crashed, etc.) And no doubt if the generalization is correct and the counterfactuals it implies are true, that is ultimately because of the *physical* structure of the machine operating in accord with the laws of physics. The lawlikeness of the macro-generalizations, insofar as they are lawlike, is parasitic on the laws of physics in a way that the laws of physics are not parasitic on anything.

The point is that this is how the special sciences work: they seek to impose a taxonomy on the physical structure of the world (the concept 'keyboard', for example, is not and cannot be reduced to the vocabulary of physics) in such a way that the objects as categorized by the taxonomy fairly reliably obey some lawlike generalizations which can be stated in terms of the taxonomy. Generalizations about how computers, or cumulus clouds, or volcanoes, or free markets behave are evidently of this sort.

Talk about 'carving nature at the joints' is just shorthand for 'finding a macrotaxonomy such that there are reasonably reliable and informative and extensive lawlike generalizations which can be stated in terms of the taxonomy', and the more reliable and informative and extensive the closer we have come to the 'joints'. Again, I claim nothing particularly novel, or astounding, about this observation.

But if the foregoing analysis is correct, we are now in a position to add something new. We have already seen that certain forms of laws, namely quasi-Newtonian laws, allow the identification of causes to be particularly simple and straightforward. So insofar as the special sciences seek to use causal locutions, it will be a *further* desideratum that the lawlike generalizations posited by the sciences be quasi-Newtonian. The special sciences, and plain common sense as well, will seek to carve up the physical world into parts which can, fairly reliably, be described as having inertial states (or inertial motions) which can be expected to obtain unless some specifiable sort of interference or interaction occurs. Or at least, those special sciences that manage to employ taxonomies with quasi-Newtonian lawlike generalizations can be expected to support particularly robust judgements about causes.

A few examples. We obviously understand computers in quasi-Newtonian terms: depending on the program being run, there is an inertial state or inertial motion, and that motion can be expected to continue unless some input comes from the keyboard or mouse, etc. The input is then an interfering factor, whose influence on the inertial state is specified by some 'laws' (the program). We similarly understand much of human biology in quasi-Newtonian terms. The inertial state of a living body is, in our usual conception of things, to remain living: that is why coroners are supposed to find a 'cause of death' to put on a death certificate. We all know, of course, that this macro-generalization is only a consequence of the clever construction of the body and a lot of constant behind-the-scenes activity: by all rights, we should rather demand multiple 'causes of life' for every day we survive. Nonetheless, the human body, in normal conditions, is sufficiently resilient that the expectation of survival from day to day is a reliable one, and the existence of the right sort of unusual circumstance immediately preceding death is typical. We do sometimes say that people just die of old age (which is obviously not a normal sort of cause) when no salient 'cause of death' exists, and our acceptance of this locution illustrates our awareness that the quasi-Newtonian generalization 'In the normal course of things (in the absence of "forces") living human bodies remain alive' is not really a law at all.

Most critically for our present purposes, we typically think of the operation of *neurons* in quasi-Newtonian form: the inertial state of a neuron is not to fire, and it departs from that state only due to impinging 'forces', namely electro-chemical signals coming from other neurons. These 'forces' come in different strengths, and can be either excitatory or inhibitory, and there is a lawlike formula that describes how the neuron will depart from its inertial state depending on the input to it. Of course, I do not mean to defend this conception as a bit of real neurophysiology, but that is how we lay people (and particularly we philosophers!) tend to think of neurons. So conceived, it is often very easy to identify the cause of the firing of a neuron. If a neuron fires and only one excitatory impulse came into it, then that impulse was the cause of firing. End of story. Counterfactuals about what would have happened had that excitatory impulse not come in (whether, for example, some *other* impulse would have come in) are simply irrelevant. Details of the larger neural net in which these two are embedded are simply irrelevant. The only thing that would *not* be irrelevant would be the discovery that the quasi-Newtonian generalization is false, e.g. because neurons sometimes spontaneously fire with no input.

The widespread use of Lewisian 'neuron diagrams' in discussions of the nature of causation is, from this point of view, both a beautiful confirmation and a deep puzzle. We like neuron diagrams because our intuitions about what is causing what are strong and robust. Those who want to analyze causation in terms of counterfactuals think that the diagrams are useful as tests of counterfactual analyses: the trick is to find some condition *stated in terms of counterfactuals about firing patterns* which picks out all and only the causes. Those conditions then tend to get very complicated, and quickly embroil one in issues like backtracking, miracles, intermediate states, etc. But it is perfectly apparent that our strong and robust intuitions in this case are not generated by fancy considerations of counterfactuals at all: they are generated by the application of quasi-Newtonian *laws* to the situation, and the counterfactuals be damned. So the puzzle is why it has not been apparent that the very diagrams used to discuss counterfactual analyses have not been recognized as clear illustrations of the wrong-headedness of the counterfactual approach.

The great advantage of the special sciences *not* being fundamental is the latitude this provides for constructing macrotaxonomies well described by quasi-Newtonian generalizations. For example, one can try to secure the reliability of a quasi-Newtonian law of inertia by *demanding* that one find an interfering factor if the inertial state changes. In a case like finding the 'cause of death' this can often be done: there is a lot of biological activity all the time, and even absences and lacks can count: one can die of starvation or suffocation. There is also latitude in carving the joints: one can shift boundaries so as to identify systems that obey more robust generalizations. This sort of boundary shifting helps explain our causal intuitions in some cases which have been discussed in the literature.

Consider the following example by Michael McDermott:

Suppose I reach out and catch a passing cricket ball. The next thing along in the ball's direction of motion was a solid brick wall. Beyond that was a window. Did my action prevent the ball hitting the window? (Did it cause the ball *not* to hit the window?) Nearly everyone's initial intuition is, 'No, because it wouldn't have hit the window irrespective of whether you had acted or not.' To this I say, 'If the wall had not been there, and I had not acted, the ball would have hit the window. So between us—me and the wall—we prevented the ball from hitting the window. *Which one* of us prevented the ball hitting the window—me or the wall (or both together)?' And nearly everyone then retracts his initial intuition and says, 'Well, it must have been your action that did it—the wall clearly contributed nothing.' (1995, p. 525)

McDermott's argument is quite convincing, but a puzzle remains. Why was nearly everyone's *initial* reaction the 'wrong' one? What were they thinking? Was it merely the falsehood of the Hume counterfactual, namely 'If you had not caught the ball it would have hit the window', that makes people judge that catching the ball is not a cause of the ball failing to hit the window? Then why does one *not* make this error (or not so easily) when the first catcher is followed by a second, infallible catcher rather than a brick wall?[4] The Hume counterfactual also fails here (if you had not caught the ball, the second catcher would have and it still would not have hit the window), but it seems

---
[4] This example is discussed by John Collins in his (2000).

clearly in this case that you are the actual cause here, and the second catcher merely an unutilized back-up. The pair of examples makes trouble for any counterfactual analysis, since the counterfactuals in each case are identical, substituting the second catcher for the wall. So if we judge causes by means of counterfactuals, our intuitions should swing the same way in both cases, but they don't.

Here is an account that makes sense of the data. In judging causes, we try to carve up the situation into systems that can be assigned inertial behavior (behavior which can be expected if nothing interferes) along with at least a partial specification of the sorts of things that can disturb the inertial behavior, analogous to the Newtonian forces that disturb inertial motion. Let us call the things that can disturb inertial behavior 'threats': they are objects or events that have the power—if they interact in the right way—to deflect the system from its inertial trajectory. We then think about how the situation will evolve by expecting inertial behavior unless there is an interaction with a threat, in which case we see how the threat will change the behavior. A threat can itself be threatened: its inertial trajectory might have it interacting with a target system, but that trajectory be deflected by something which interferes with *it*. This is what we mean by *neutralizing* a threat, or by *preventing* an event.

Now what counts as inertial behavior, and what counts as a threat, depends on how we carve the situation up into systems. If we consider a normal window on its own, its inertial behavior is to remain unbroken (something must cause it to shatter, but nothing causes it not to if nothing interferes with it). The sorts of thing that count as threats are objects with sufficient mass and hardness and (relative) speed: a hurled rock is a threat to a window but a lofted marshmallow is not. The cricket ball in the example is obviously a threat to break the window (although the case just deals with hitting the window, the same point holds). The cricket ball is furthermore in a state of motion such that its (Newtonian) inertial trajectory has it hit the window, so for the threat to be neutralized, something must act to deflect the ball. This is clearly the first catcher in both cases, and the presence of the wall or second catcher never comes into the analysis.

But let us now carve up the situation a bit differently. Let us call the window + brick wall system a 'protected window'. The inertial state of the window in a protected window system is to remain unbroken, and indeed to remain *untouched*. In order to count as a threat to this state, an object must be able to *penetrate* (or otherwise *circumvent*) the wall, at least if it is approaching from the direction protected by the wall. Thought of this way, the cricket ball is *not* a threat to the inertial behavior of the window: only something like an artillery shell would be. So in the given situation, there are no threats to the window at all, and a fortiori no threats which are neutralized, and so no prevention: *nothing* causes the protected window not to be hit since that is its inertial state and nothing with the power to disturb the inertial state threatened.

In sum, carved up one way (window plus wall plus catcher plus ball) we get systems governed by quasi-Newtonian generalizations which yield the judgement that the catcher prevented the window being hit, carved up another way (protected window plus catcher plus ball) we get systems governed by quasi-Newtonian generalizations which yield the judgement that the window was never threatened with being hit, so nothing had to prevent it. And the interesting thing is that *both systematizations, together with the corresponding quasi-Newtonian generalizations, yield the same counterfactuals.* So no counterfactual analysis of causation has the resources to account for the disparity of judgements: carving up the world differently can give different (special science) *laws*, governing different *systems*, but it should not give different truth values to counterfactuals.

If this analysis is correct, McDermott has played something of a trick on his test subjects. Their original intuition that the catcher does not prevent the window being hit is perfectly correct, relative to conceptualizing the situation as containing a protected window. McDermott then asks a series of questions which essentially require that one reconceive the situation, carving it instead into a window, a ball, and a compound wall + catcher system. So thought of, the ball is clearly a threat to the window, the threat needs to be neutralized (or deflected) to prevent the window being hit, and the wall + catcher system does the deflecting. The last step is to give the credit to either the wall or the catcher or both, and here the catcher clearly wins. All of this is perfectly legitimate: it is just no *more* legitimate than lumping the window and wall together and judging that nothing prevented the hit.

The window plus two catchers example tends to confirm the analysis: here the natural tendency would be to regard the two catchers as equally autonomous systems: it would be odd to carve up the situation so that the window plus second catcher is regarded as a single system (although one could tell stories which would promote this). So our intuitions go differently from the wall case, even though the corresponding counterfactuals are the same.

The only remaining question is why typical subjects would have the tendency to regard the situation as containing a protected window, rather than in terms of a regular window, a threatening ball, and a pair of potential neutralizers. Lots of hypotheses come to mind, and the only way to test them would be to try many more sorts of examples. Windows and walls are obviously more similar than catchers and walls, since windows and walls are inert bits of architecture. It is *simpler* to keep track of things using the 'protected window' systematization, since there are fewer threats and hence fewer threats that need to be tracked. The 'protected window' systematization also has the advantage of yielding (in this case, at least) causal judgements which agree with the Hume counterfactual: the catcher does not prevent the window being hit, and the window would not have been hit even if the catcher had not acted. (The last two considerations are held in common with the 'two catcher' case, where our intuitions go the other way, but in that case there is positive pressure to group the catchers together.)

A thought experiment recommends itself: imagine various ways for the wall to metamorphose into a second catcher and track when one's causal intuitions flip. But that is a task for another paper.

Alternative taxonomies need not be at the same degree of resolution as in the foregoing example. Ned Hall has argued in (2004) that cases of double prevention (preventing an event which would have prevented another event) are typically not regarded as causation (in the sense of production). Jonathan Schaffer (2000) has then pointed out that guns fire by double prevention, but one is not tempted to conclude that pulling the trigger of a gun is not a cause of the firing. But surely what guide our intuitions in this case are lawlike macro-generalizations in quasi-Newtonian form. The 'inertial law' for guns is just: a gun doesn't fire if nothing acts on it (e.g. pulls the trigger, jars it, etc.). *Anything which regularly results in a gun firing (particularly pulling the trigger) counts as a cause that changes the inertial state, no matter how the trick is done at the micro level. This judgement is not reversed even if one decides that the right thing to say at the micro level is that pulling the trigger does not produce the firing.* If different taxonomies can allow for different lawlike generalizations, and hence different causal judgements, we have all the makings for interminable philosophical disputes, since causal judgements can be reversed simply by changing taxonomies, as McDermott's example illustrates.

5. REMOTE CAUSATION

The foregoing account has been concerned with the analysis of what we may call *proximate* or *immediate* causation: where a situation is conceptualized as governed by quasi-Newtonian laws, the laws will specify what counts as an inertial state, and therefore what counts as a deviation from an inertial state (a first-class effect), and also what sorts of things or events (causes) bring about such deviations. The interfering factor is then the proximate cause of the deviation, as in the particle collision example.

This analysis does not, however, solve the problem of *remote* causation. We commonly identify events or actions as causes that are not the proximate causes of their effects: the loss of the horseshoe nail ultimately causes the loss of the battle not directly but by a long chain of intermediates. The loss of the horseshoe nail is the proximate cause only of the loss of the horseshoe.

Prima facie, the use of quasi-Newtonian laws appears to be of some help in this regard. If the laws have the form of specifying inertial states and interfering factors, then any concrete situation can be represented by what we may call an *interaction diagram*, a graph depicting inertial motions of objects as straight lines and divergences from inertial motion as always due to the interaction with some interfering element. One might think, for example, of Feynman diagrams as examples of the form. Note that if all interactions are local, or by contact, then interaction diagrams will look like space-time diagrams, and the lines will represent continuous processes in space-time, but this is not essential to the form: if there is action-at-a-distance, then two lines can intersect on an interaction diagram even if the objects they represent never come near each other in space-time.

Interaction diagrams supply one very simple method for identifying remote causes: trace the diagram backward from an event, and every node one comes to counts as a cause. But this method will not, in many cases, accord with our intuitions. Being in the interaction diagram for an event may well be a necessary condition for being a cause, but it is unlikely to be sufficient. Interaction diagrams will include events that we consider not to be causes (in the usual sense) but failed attempts at prevention, such as the futile course of chemotherapy that fails to stem the cancer. Interaction diagrams also stretch back indefinitely in time, to events too remote to be commonly considered causes. They are a bit more parsimonious than the entire back light-cone of an event, but not that much.

This sort of situation is, of course, the siren song for the analytic philosopher. Perhaps we can identify commonsense remote causes by use of an interaction diagram plus some further condition (the Hume counterfactual, for example). But the avalanche of counter-examples to the many theories of causation that have been floated ought to give us pause. Perhaps there is some reason that the analysis of remote causation has proved so difficult.

So let's step back a moment and ask what *purpose* there is to identifying remote causes in the sort of situation we are contemplating, i.e. where we know the laws, the immediate causes of events, and the complete interaction diagram. What more could we want? What do we gain by trying to distinguish events that count as remote causes from others that are not?

One thing we might have is a practical concern for prediction and/or control. We might like to know how we could have prevented a certain event, or whether a similar event is likely to occur in future situations that are, in some specified respects, similar. But in that case, all we would really care about is the Hume counterfactual—and that, as we know by now, is neither sufficient nor necessary for causation. So what other aim might we have?

In some circumstances, we are interested in remote causes because we wish to assign *responsibility* for an event, e.g. when offering praise, or assigning blame, or distributing rewards, or meting out punishments. In these circumstances, identifying a remote cause is often tantamount to establishing a responsible agent, and many of our intuitions about remote causes have been tutored by our standards of culpability. If this is correct, then we might make some progress by reflecting on such standards.

The game of basketball provides an example where rules for assigning credit as a remote cause have been made reasonably explicit. Consider this case: a pass is thrown down court. The center leaps and catches the ball, then passes it to Forward A, who dunks the ball. Forward A gets credit for the field goal, and the center gets an assist: he counts as a remote cause of the points being scored. Note that these attributions do not change even if it is also true that the pass was intended for Forward B, who was standing behind the center, and that had the center not touched the ball, Forward B would have caught the pass and scored, or would have passed it to A. Even if the Hume counterfactual does not hold for the center (even had he not caught the ball, the points would have been scored), the center counts, unproblematically, as a cause.

Now consider an apparently analogous case. John enters a car dealership, unsure whether he will buy a car. He is met by an Official Greeter, who directs him to Salesman A. Salesman A makes the pitch, and convinces John to buy the car. Had the Greeter not intercepted him, John would have run into Salesman B, who also would have convinced him to buy the car (or who, perhaps, would have directed him to Salesman A). In this case, the Greeter cannot, intuitively, claim credit for the sale: he did not, even remotely, cause

Causation, Counterfactuals, the Third Factor   167

John to buy the car. He was a cause of the cause, a cause of John's hearing Salesman A's pitch (at least in the scenario where the alternative is a pitch from Salesman B), but we are not inclined to accept transitivity here.

Doubtless there are perfectly good reasons for the difference in practices for assigning credit in these cases: *typically*, in a basketball game, had the assist not occurred points would not have been scored, so one wants to recognize and encourage those who give assists. But *typically*, in the sort of situation described in the car dealership, whether a sale is made does not depend on the actions of the Greeter. But what is typical does not affect the individual case: we can make the counterfactual structure of this *particular* pair of examples as analogous as we like (the schematic 'neuron diagrams' for the scenarios can be made identical) without changing our views about who deserves credit and who does not. If so, then *no* generic account of remote causation couched in terms of interaction diagrams or counterfactual structure will always yield intuitively acceptable results: in different contexts, our intuitions will pick out different nodes on the same diagram as remote causes. In these cases, the definition of a remote cause is, as Hume would put it, 'drawn from circumstances foreign to the cause', i.e. from statistical features of other cases that are regarded as similar. As such, the standards violate the desideratum that causation be intrinsic to the particular relations between cause and effect.

If standards for identifying remote causes vary from context to context, and, in particular, if they depend on statistical generalities about *types* of situations rather than just on the particular details of a single situation, then the project of providing an 'analysis' of remote causation is a hopeless task. We might usefully try to articulate the standards in use in some particular context, but no generic account in terms of interaction diagrams or counterfactual connections will accord with all of our strongly felt intuitions.

6. THE METAPHYSICS OF CAUSATION

If the foregoing analysis is correct, then (1) what causes what depends on the laws which govern a situation, (2) judgements of causation are particularly easy if the laws have quasi-Newtonian form, (3) everyday judgements ('intuitions') about causation are based not on beliefs about the only completely objective laws there are (namely physical laws) but rather more or less precise and reliable and accurate lawlike generalizations, (4) the same situation can be brought under different sets of such generalizations by being conceptualized differently, and those sets may yield different causal judgements even though they agree on all the relevant counterfactuals.

To what extent, then, is causation itself 'objective' or 'real'? At the level of everyday intuition, the freedom to differently conceptualize a situation implies that one's causal judgements may not be dictated by the complete physical situation *per se*. Further, the lawlike generalizations appropriate to the conceptualization can be criticized on objective grounds: they could be more or less accurate or clear or reliable. If, for example, windows sometimes spontaneously shatter (due to, say, quantum fluctuations), then the reasoning that the ball made it shatter because the ball was the only thing to hit it (and its inertial state is to remain unbroken unless something hits it) is no longer completely trustworthy, the less so the more often spontaneous shattering occurs. Lawlike generalizations are also supposed to support counterfactuals, but those counterfactuals must ultimately be underwritten by physical law, so a close examination of the physics of any individual situation could undercut the macro-generalization applying to that case. The quest for greater scope, precision, and reliability of generalizations tends to force one to more precise micro-analysis, ultimately ending in the laws of physics, which brook no exceptions at all.

If the laws of physics turn out to be quasi-Newtonian, then there could be a fairly rich objective causal structure at the fundamental level. But if, as seems more likely, the laws of physics are not quasi-Newtonian, then there may be little more to say about physical causation than that the entire back light-cone of an event (or even the entire antecedent state of the world in some preferred frame) is the cause of the event, that being the minimum information from which, together with the laws of physics, the event can be predicted. Quasi-Newtonian structure allows one to differentiate the 'merely inertial' part of the causal history of an event from the divergences from inertial states that are paradigmatic effects, but without that structure it may be impossible to make a principled distinction within the complete nomically sufficient antecedent state. In that case, causation at the purely physical level would be rather uninteresting: if all one can say is that each event is caused by the state of its complete back light-cone, then there is little point in repeating it. None of this, of course, is of much interest to physics *per se*, which can get along quite well with just the laws and without any causal locutions.

There is much more that needs to be said about the role laws, or lawlike generalizations, play in the truth conditions of causal claims. The proposal

I have made is more of a sketch than a theory, and has all the resources of vagueness to help in addressing counter-examples and hard cases. Perhaps there is no adequate way to make the general picture precise. But the counterfactual approach to causation has had a good long run, and has not provided simple and convincing responses to the problem cases it has faced. Perhaps it is time to try something new.

6

The Whole Ball of Wax

In the introduction to the second volume of his collected papers, David Lewis remarked:

> Many of the papers, here and in volume I, seem to me in hindsight to fall into place within a prolonged campaign on behalf of the thesis I call 'Humean supervenience'. Explicit discussion of that thesis appears only in 'A Subjectivist's Guide to Objective Chance'; but it motivates much of the book.
>
> Humean supervenience is named in honor of the greater [*sic*] denier of necessary connections. It is the doctrine that all there is to the world is a vast mosaic of local matters of fact, just one little thing and then another. (But it is no part of the thesis that these local matters of fact are mental.) We have geometry: a system of external relations of spatio-temporal distance between points. Maybe points of spacetime itself, maybe point-sized bits of matter or aether fields, maybe both. And at those points we have local qualities: perfectly natural intrinsic properties which need nothing bigger than a point at which to be instantiated. For short: we have an arrangement of qualities. And that is all. There is no difference without difference in the arrangement of qualities. All else supervenes on that. (Lewis 1986a, pp. ix–x)

It would be disingenuous of me to claim that only in retrospect have I realized that the papers presented here, written over a span of more than fifteen years, all fall in a prolonged campaign against Humean supervenience: after all, one is entitled 'Why Be Humean?' But there is a certain coherence and commonality of purpose in these papers which was not evident to me as they were being written. There a method to this madness, which I only came to appreciate because of some astute observations of Barry Loewer, who has spent at least as much effort refining and defending Humean supervenience as I have spent trying to undermine it. So it seems appropriate to briefly state explicitly the foundational picture implicit in the essays that have come before.

The Humean picture of ontology, as Lewis understands it, is founded on the notion of the Humean Mosaic: a collection of local qualities structured into a unified object by external spatio-temporal relations. At base, Lewis's

Humean believes that this is all there is: it is not merely, as Lewis says, that everything else *supervenes* on the Mosaic, but rather that anything that exists at all is just a *feature* or *element* or *generic property* of the Mosaic. If one wants to say there are laws, for example, then what the laws are is simply a matter of how the Mosaic is structured: the philosophical problem is to specify, as clearly as possible, which features of the Mosaic *constitute* the laws having a certain form. If there is a direction of time, then that too is just a feature of the Mosaic: the philosophical problem is to specify what features the Mosaic must have in order that time have a direction. These sorts of philosophical analyses then count as reductions of ontology supposing (as Lewis does suppose) that the basic elements of the Mosaic, in themselves, are not ontologically dependent on either the laws or the direction of time. (So it had better not be a basic Humean particular fact that the laws in some region are such-and-such, or that the direction of time at some event points a particular way.) The elements of Lewis's Mosaic must have intrinsic characters that depend on the existence of nothing else in the universe, and these elements must be related by some purely external relations, so that all of ontology can ultimately be resolved into the set of pointlike elements and their arrangement.

Lewis identifies Hume as the great denier of necessary connections, but one should take some care about the sort of necessity Hume worried about, and, particularly, whether those worries ought to inform contemporary metaphysics. One concern that Hume had was about the sort of necessity implicit in the notion of a *causal* connection. And Hume's concern on this point had straightforward empiricist roots: it is no part of our initial *experience* of a pair of events that we *sense* or *feel* a necessary connection between them. Hume's question, then, was from what simple *impression* the idea of necessary connection—and hence causation—could be derived. But from a contemporary perspective, all of this concern is misplaced. Suppose, for example, that we have secured facts about what the laws of nature are (either by a reductive analysis or, as I would prefer, as a basic posit). Then there are models of those laws: physical situations that are consistent with the laws. And we can treat those models as possible worlds in the usual way, and thereby define a notion of nomic possibility and necessity. And that is all the necessity that causation needs. We have provided clear truth conditions for claims about a certain sort of necessity without having to produce an impression to be the original of the idea, as Hume would have wanted.

The issue, then, is not the definition of terms like 'causation' or 'necessity': the issue is what the consequences are of taking the Humean Mosaic as ontological bedrock as opposed, say, to taking the laws of nature and the direction of time as ontological bedrock. And it is here that Loewer made the key observation.

I hold an unpopular view about laws of nature: they ought to be accepted as ontological primitives. I hold another unpopular view about the passage of time: it too is an ontological primitive, which accounts for the basic distinction between what is to the future of an event and what is to its past. The appeal of these two unpopular views is, for me at least, connected. It was perhaps already clear when I wrote 'A Modest Proposal ...' that the issue of time and the issue of natural laws were deeply intertwined: I noted in that essay that the fundamental laws of nature appear to be laws of *temporal evolution*: they specify how the state of the universe will, or might, evolve from a given initial state. Therefore investing these sorts of laws with an irreducible ontological status comports well with investing the direction of time with that same status. What Loewer pointed out—and what seems to me correct—is that it would be much harder to understand the advantages of making one of these posits without the other than it is to understand why it makes sense to make them both together. And the fundamental issue here is one of *explanation*.

If one is a Humean, then the Humean Mosaic itself appears to admit of no further explanation. Since it is the ontological bedrock in terms of which all other existent things are to be explicated, none of these further things can really *account for* the structure of the Mosaic itself. This complaint has been long voiced, commonly as an objection to any Humean account of laws. If the laws are nothing but generic features of the Humean Mosaic, then there is a sense in which one cannot appeal to those very laws to *explain* the particular features of the Mosaic itself: the laws are what they are in virtue of the Mosaic rather than vice versa.

One problem making this argument out is that it seems to run afoul of any purely *syntactic* theory of explanation. As an obvious example, the D-N model of explanation would suggest that all that matters for the explanatory power of laws is their propositional content. Following the terminology of Loewer (1996), let us call the output of the Mill–Ramsey–Lewis mechanism the *L-Laws*. It is quite plausible that any laws of physics will be L-Laws: members of the simplest, strongest accounts of the global state of the universe. These L-Laws can be plugged into the D-N model just as well as any other sort of generalization, and so, if the D-N model is correct, can provide explanations.

Or, to take a claim made by Loewer, the L-Laws can explain features of the Mosaic by *unifying* them: by showing commonalities of structure among various distinct events or processes or regions of space-time (1996, p. 113). So the general question of whether Humean laws can explain particular events must evidently be held hostage to a systematic account of explanation.

But there is an important point that can be made even without such a systematic account. There may be some sense in which Humean laws explain particular events, but there is a clear sense, I think, in which they cannot. This sense is closely connected to examples that have long illustrated the flaws in the D-N model.

Consider the classic example used against the D-N model: Bromberger's flagpole. The particular physical circumstances—height of the flagpole, position of the Sun, meteorological conditions, etc.—together with the laws of optics imply the length of the flagpole's shadow and, we think, also *explain* that length. But the same particular details, omitting the height of the flagpole and replacing it with the length of the shadow, together with the laws of optics, equally imply the height of the flagpole. For all that, our strong intuition is that while this *establishes* or *proves* what the height is, it does not, in any sense at all, *explain* the height. If so, then the D-N structure is not sufficient to provide the relevant sort of explanation, nor would considerations of, say, unification seem to help the situation.

It is, at one level, very easy to see what is going on in the flagpole example: the account of the shadow in terms of the attendant circumstances and the laws of optics is a complete account of how the shadow is *produced*, and hence an explanation of all of its particular features, including its length. The inference from the attendant circumstances (including the shadow's length) to the height of the flagpole, in contrast, tells us absolutely nothing at all about how the flagpole was produced, and therefore cannot account for *any* of its particular features, even though it can serve to indicate what those features are.[1]

So there is one sort of explanation we recognize, an explanation that consists in an account of *how an item (or event) was produced*, which does not

---

[1] In an amusing story in his (1980), Bas van Fraassen appears to want to construct a scenario in which the length of the shadow of a tower does, indeed, explain the tower's height (pp. 132–4). But evidently it is rather *the intention that the tower cast a certain sort of shadow* (which intention pre-dates and plays a causal role in the construction), rather than any feature of the shadow itself, that does the explanatory work.

easily fit into either the D-N or unifying model of explanation. It is this sort of explanation we can provide of the shadow, but not of the flagpole itself, in the situation under consideration.

Are such accounts of production *causal* explanations? Often they are, and we might even make them all so by stipulation. Ned Hall (2004) has persuasively argued that there are two quite distinct concepts of causation: *dependence* (basically counterfactual dependence of a certain sort) and *production*, and this seems like a reasonable way to proceed. But having said that, it also seems notable that accounts of the production of something need not be deterministic (and hence, in that sense, causal): the production of an item might proceed via various stochastic events that could have turned out differently. David Lewis has also pointed out that these sorts of explanation need not make the explanandum seem more *likely* or *to-be-expected*:

> Explanations give causal or nomological information. That information often does make the explanandum less surprising, but it may make it more surprising, or may leave it about as surprising as before. Suppose you check into a hotel room, and there you find a new-looking pack of cards. They turn out to be ordered neatly: they go from the ace to king of clubs, then ace to king of diamonds, then ace to king of hearts, then ace to king of spades. Not surprising—maybe it's a brand new deck, or maybe whoever left them had won at solitaire. Not so. What's true is that they got into that order by being well and fairly shuffled. The explanation, if known, would make the explanandum much more surprising than it was before. (Lewis 1986b, p. 133)

This sense of 'explanation'—and it is a plainly acceptable sense of the word in English—is just an account of how 'they got into that order': it is a *history of the production of the state*. So the question before us is what, if anything, would count as a similar explanation of the Humean Mosaic itself, an explanation of how it got into the state it is in. (Or, more precisely, what would count as a productive explanation of the *total physical state of the universe*, which may or may not properly be regarded as a Humean Mosaic. More on this below.)

If we believe in the laws of nature—the Fundamental Laws of Temporal Evolution, as I call them in 'A Modest Proposal...'—as ontological primitives, and we believe in the direction of time as an ontological primitive, then it is clear how such an account can be given. The universe started out in some particular initial state. The laws of temporal evolution operate, whether deterministically or stochastically, from that initial state to generate or produce later states. And the sum total of all the states so produced *is* the Humean Mosaic.

This counts as an explanation exactly because the explanans (namely the initial state, or the state up to some time, and the laws) are ontologically distinct from the explanandum (namely the rest of the Mosaic). The laws can *operate* to *produce* the rest of the Mosaic exactly because their existence does not ontologically depend on the Mosaic. If it did (as the Humean would have it) then they could not play this sort of role in producing the Mosaic, and hence could not play any role in this sort of explanation of the Mosaic.

This is not simply a D-N explanation because the notion of production is asymmetric while logical implication may well be symmetric. If the laws are of the right form, then the initial state can be *inferred* from the final state (or from some later state) plus the laws, but this does not constitute an account of how the initial state came about. The basic temporal asymmetry of past-to-future underlies the very notion of production itself, so that without it there can be no production. Hence, once again, any account of the direction of time that makes it *ontologically* dependent on the global structure of the Mosaic cannot provide this sort of productive account of the Mosaic itself.

Furthermore, I think this sort of productive explanation cannot abide closed timelike curves or backward time travel (even if there is a unique, primitive direction of time). The explanans must be ontologically independent of the explanandum, the producers of that which is produced, so nothing can be the producer of itself. The ancient philosophical prohibition against self-caused or self-created items reflects the constraints on this sort of productive explanation. Production is clearly transitive, so production-in-a-circle would imply self-production, but an item cannot be the *ontological* ground of its own production. This means that the existence of closed timelike curves would imply the non-existence of this sort of productive explanation, and might suggest that a Humean account is the strongest that can be had. But there is, as yet, no evidence at all for the existence of closed timelike curves, and arguments that present physics suggests the *possibility* of such curves (but not their *actuality*) are question-begging. If the account of the passage of time I am defending entails that time cannot pass in a circle, and if the structure of *actual* time is consistent with this theory, then the conclusion from this perspective is that mathematical models containing representations of closed timelike curves do not represent possible structures of time itself, and hence do not represent physical possibilities. (See the Epilogue, 'A Remark on the Method of Metaphysics', this volume, for a more extensive discussion.)

The temporal asymmetry of productive explanations not only coheres with the intuitive view that the passage of time is an inherently asymmetrical matter, it also coheres with a widely noted form of explanation of other manifest asymmetries, such as the asymmetry of entropy in the universe. The *full* productive explanation of the Humean Mosaic postulates an initial state of the universe and dynamical laws, and, possibly, the outcomes of fundamentally stochastic evolution if the laws are not deterministic. This account would encompass every particular detail of the Humean Mosaic. But there are also generic characteristics of the Mosaic that appear to admit of more generic explanations. As David Albert has argued, the manifest asymmetry in our knowledge of the past and the future, and the manifest asymmetries of thermodynamics, are all explained if one supplements (even time reversible) dynamical laws with the *Past Hypothesis* 'which is that the world first came into being in whatever particular low-entropy highly condensed big-bang sort of macrocondition it is that the normal inferential procedures of cosmology will eventually present to us' (Albert 2000, p. 96). Note that the Past Hypothesis is generic in that it specifies only the *macro*condition of the initial state, whereas the exact Humean Mosaic will depend on the microcondition. The Past Hypothesis, together with the postulate that observable behavior in the universe will be *typical* for this macrocondition, allows us to explain both thermodynamics and the manifest asymmetries of epistemic access.[2]

I consider this a form of generic productive explanation, showing how certain features of the world are understandable if the world is produced in a certain way from a certain kind of initial state. This sort of explanation takes the term *initial* quite seriously: the initial state temporally precedes the *explananda*, which can be seen to *arise* from it (by means of the operation of the law). And on account of this, I cannot agree with another claim made by Albert: that the same structure will explain the overall asymmetry of causation (namely that causes precede their effects), but only as a matter of *preponderance*: causes *mostly* or *typically* precede their effects, but some effects precede their causes. This is the case with entropy (it *mostly* increases, but can briefly decrease through lawful fluctuations) and with knowledge (we have *more* detailed information about the past than the future), but there are no known or recognized instances of causation running from future to past. If the relevant notion of causation is a form or aspect of production, and

---

[2] There is a somewhat technical issue about the precise form of the second postulate needed for these explanations. Albert uses what he calls a 'Statistical Postulate', which employs a specific probability measure over the microconditions compatible with the macrocondition, but there is a logically weaker notion of *typicality* that can underwrite the same explanations. The details are discussed in my 'What Could Be Objective about Probability?' (2007).

if production involves an inherent asymmetry in time direction, then one gets this result as a metaphysical necessity. On Albert's accounting, there is nothing *intrinsically* time directed about causation, but the Past Hypothesis should entail that any backward causation is not the sort that we could know about, or use for our own ends. Clearly there is no *empirical* argument that can decide this issue.

Nor could considerations of parsimony decide the issue: both Albert and I provide metaphysical analyses of causation in terms of more primitive items: in his case, the Humean Mosaic (with attendant Humean Laws) plus a semantics of counterfactuals and a counterfactual analysis of causation; in my case, in terms of primitive law and the direction of time. He gets a causal asymmetry that holds only for the most part, and depends on the Past Hypothesis; I get a notion of production built on the foundational temporal asymmetry that would obtain even if the world were always in thermal equilibrium (even then, later states would arise out of earlier ones). His analysis contravenes common intuition by implying that some of our present actions can cause or influence events in the past, mine accords with common intuition. Albert, of course, also thinks that his analysis will *account for* the common intuition, since instances of backward causation must necessarily reside behind an epistemic veil, but this claim itself gets embroiled in the issue under debate: if one wants to account for a present, commonly held intuition by explicating *how it came about* (via evolutionary psychology, for example), then we get right back to the question of whether the Humean can explain how *anything* came about in the sense that the non-Humean can.

It is part and parcel of the sort of productive explanation that I have in mind that any appeal to boundary conditions must be an appeal to the initial boundary, the earliest boundary, rather than a future boundary. It was long considered a hallmark of the scientific revolution that *teleological* explanations were abandoned: no particular event or process is to be explained by appeal to the *effects* it will have or the particular end state to which it leads. Given a fundamental distinction between the past-to-future and the future-to-past direction, then, any further postulates must be about the past state of the universe. The Humean, by contrast, can make no clear sense of the prohibition. A constraint on a boundary is just that, and is to be accepted as itself lawlike if it helps to compactly convey information about the Mosaic. Whether there is any direction of time, or whether the rest of the Mosaic is produced *from* the boundary, or whether, given a direction of time, the boundary in question lies to the future or the past, are all irrelevant questions.

There is a further distinction to be made between *fundamental* and *pragmatic* productive explanations. A fundamental explanation makes use of a Fundamental Law of Temporal Evolution, and the *input* into the explanation—the state from which the production proceeds—must be whatever the Laws require to specify the evolution (whether deterministically or stochastically). Typically, this will be a very large set of data: data for the exact entire physical state of the universe, or perhaps data for a hypersurface that cuts through the complete back light-cone of an event. The laws need all of these given data in order that the evolution of the physical state be determined. Everyday productive explanations, in contrast, confine themselves to small subsets of these data: the present scar on my leg was produced by a particular cut many years ago, not by absolutely everything in the past light-cone of the present scar. Much philosophical attention has been paid to the principles of pragmatic causal explanation, to the way we sort the *causes* from the merely *attendant circumstances*, and there is some conceptual interest in this question,[3] but it seems completely irrelevant to matters of fundamental ontology. The fundamental laws of nature make no such distinction: they need to know the entire relevant past state, in all detail, in order to know how to go on.

The non-Humean, then, has available a distinctive sort of explanation of the total physical state of the universe that the Humean must, of necessity, forswear. This explanation is bought at an ontological price, which includes both primitivism about laws of temporal evolution and primitivism about the direction of the passage of time. The extra explanatory resources should not be surprising since the non-Humean has a larger ontology to work with: the complete physical state of the whole universe through all time as well as non-supervenient laws and a direction of time. The Humean ontology may not be strictly a subset of mine, since I have no obligation to regard the complete physical state of the universe as a Humean Mosaic (i.e. as a collection of local, non-nomic states of affairs knit together by external relations), but still in any intuitive sense my ontology is larger than the Humean's. One *ought* to be able to buy something with this extra ontology, and a productive account of the complete physical state itself is the most obvious item on offer.

So we have two competing accounts of the basic ontological structure of the universe, with different primitives and different sorts of available

---

[3] One account of how to make this distinction in the case of quasi-Newtonian laws can be found in 'Causation, Counterfactuals, and the Third Factor', Chapter 5, this volume.

explanations. The question is: by what methodological principle ought one to decide between these ontologies?

There are two commonly invoked methodological principles that occur in these sorts of discussions, and prima facie they appear to pull in opposite directions. One is Ockham's Razor and the other is the principle of Inference to the Best Explanation. Using the first, one might (rather glibly) argue that the Humean account is to be preferred since it has a smaller ontology, with fewer ontological primitives. Using the second, one might argue (equally glibly) that the non-Humean ontology ought to be preferred since it provides a better explanation of the total physical state (the Humean Mosaic) than the Humean view does, since the Humean provides, in the relevant sense, no explanation at all. In fact, I think that neither of these principles has a straightforward application in this case, and further, when properly understood, they amount to the same methodological principle.

Let's start with the Razor. I can think of no reason to suppose that smaller ontologies are *per se* preferable to larger ones, when comparisons of size are not under dispute. Even when the ontology of one theory is a proper subset of the ontology of the other, no *logical* principle suggests that the *ontological claims* of the one are better supported than the ontological claims of the other, for those ontological claims will include *denials* as well as posits. For example, when comparing a field-theoretic account of classical electromagnetism with an action-at-a-distance account, the ontology of the latter can be made strictly a subset of the ontology of the former, but that in itself gives the action-at-a-distance version no more credibility. Indeed, in this situation denying the existence of the fields seems much less plausible than affirming their existence: the fields allow the laws to be simplified, secure various conservation laws, and so on. So the Razor cannot be a principle that renders a judgement simply on the basis of the size of an ontology.

There are particular cases where we must decide between competing theories and where the Razor can rightly be invoked. One is where an intuitively simpler set of postulates, taken as a whole, can account for the same phenomena as a more complicated or contrived set. The question is how to make sense of the relevant notions 'simpler', 'complicated', and 'contrived'. If all of the lights in my house suddenly go out, it might be that they all burned out at the same moment by coincidence or that something has interrupted the electricity to all of them. Any reasonable person would embrace the latter explanation and reject the former, and rightly so. But why?

In a certain sense, the first explanation posits more novel items (i.e. items, unlike the lights themselves, that I don't already believe in) than the latter: multiple burnings-out as opposed to a single central disruption. But sheer counting can't be all that matters: a proposed explanation of a disease that posits the interaction of a virus and a particular environmental factor is not, *per se*, less plausible than one that just posits the virus. The real issue here is not mere numbers but *co-ordination*: it would be quite an unexpected and atypical *coincidence* for all of the bulbs to burn out at once. The first explanation requires, by its own lights, a very special choice of the variable parameters to get the noted effect (all the lights going out at once) while the latter doesn't. It is easy to see how, e.g. a Bayesian can give a standard confirmation-theoretical account of why the second explanation deserves more credence than the first. So we have the following rough-and-ready explanation of the appeal of the Razor: more complicated explanations, explanations with more 'moving parts', as it were, will often have more independent parameters that need to be specified. So it would not be surprising if recovering a particular effect required some fine-tuning of the parameters relative to one another. Absent any further principles constraining the parameters, this fine-tuning will end up just being fortuitous: a matter of chance. But then, by the theory's own lights, the effect would be assigned a low a priori probability. And ceteris paribus such a theory will get less of a confirmational boost than one that manages to assign a higher probability to the effect.

But having laid out this account of the proper application of the Razor, we see that exactly the same account can be given of Inference to the Best Explanation. Explanations are supposed to be better insofar as they are simpler, but the exact content of 'simpler' is not evident. If simplicity were just a matter of the amount of postulated ontology, then we would immediately get back our rejected version of the Razor. It is more defensible to tie simplicity also to *lack of fortuitousness* or *reduction of required coincidence*. Given a reasonably robust background against which degrees of coincidence can be evaluated, we again get the straightforward Bayesian result that ceteris paribus an account that posits less coincidence (such as the 'unused pack' and 'won at solitaire' hypotheses discussed by Lewis above) will be more plausible that an account that posits more (such as 'well and fairly shuffled').

But the key observation is now this: judgements about relative degrees of coincidence can only be cleanly made within a common background theory. It is part of our common general account of the world that packs of cards can be unused, used to play solitaire, and be well and fairly shuffled. We also have a sense of how common each of these conditions is, and what the conditional probability for the pack to be well ordered at the end in each scenario is. Being unused or having won at solitaire is a better explanation of the well-ordering exactly because the probability for the well-ordering is so much higher on each hypothesis. Given that the priors for the three hypotheses (before looking at the cards) are of the same order of magnitude, the posteriors for the 'better explanations' will be much higher.

But judgements of degree of coincidence *between* fundamentally different ontological pictures are a much trickier business. The non-Humean may well be tempted to argue thus: if there were no laws *operating* to *produce* the Mosaic, how likely is it that the Mosaic would exhibit any regularities at all? After all, if one just *randomly assigned* the local quantities to different positions in the Mosaic, there would probably be no discernible pattern in the result. But this is really a cheat: it is not seriously considering the hypothesis that the Mosaic was not produced at all, it is rather considering the hypothesis that it was produced by a random process. Such a process *would* assign a straightforward chance to getting a patterned result, but the Humean postulates no such process. Lacking enough background ontology to support the assessment of a probability, the notion of a *coincidence* cannot take hold here.[4]

So both Ockham's Razor and Inference to the Best Explanation are of no use, I think, in resolving the dispute between the Humean and the non-Humean. Why then am I so resolutely non-Humean? Why think that laws of nature and the direction of time have a primitive ontological status rather than being subject to a Humean analysis?

The first motivation of these papers was simply to point out the defensibility of these claims. In the case of laws of nature, naive scientific practice takes the fundamental laws as objects of enquiry but not analysis: it is assumed that there are such things, and our job is to find out what they are. The notion that the laws themselves have a derivative ontological status, that they are nothing but generic features of some Humean Mosaic, is not to be found in science itself. As a philosopher, I am inclined to try to take scientific practice at face value, and in this case I could see no good reason not to. Hume's own reasons, based in a discredited account of concept formation and semantics,

---

[4] Similar remarks apply to the so-called fine-tuning arguments put forward in favor of the hypothesis that the universe was designed. There is no background theory of the 'origin' of the constants of nature that would allow for any justifiable judgement about how 'likely' they are to take certain values. One sometimes falls back on metaphors of God 'picking' the values of the constants, but the non-metaphorical content of this image is empty.

certainly no longer provide a motivation. In the case of the passage of time, the situation is not so straightforward: many physicists (but not all) believe that acceptance of a Relativistic account of space-time, or acceptance of time-reversible dynamical laws, is inconsistent with the notion that time passes. Here I have tried to point out why the arguments to these conclusions do not go through. Certainly the sorts of scientific explanations that would really be inconsistent with the passage of time—explanations of the present by appeal to future conditions, or explanations that require temporal loops—do not actually exist. Present physics, as practiced, can accept the passage of time as a fundamental asymmetry without explanatory loss.

Let's call the idea that both the laws of physics (as laws of temporal evolution) and the direction of time are ontological primitives *Maudlin's Non-Humean Package*. According to this package, the total state of the universe is, in a certain sense, derivative: it is the product of the operation of the laws on the initial state. The total state need not, of course, be a *logical consequence* of the laws and the initial state: the laws might, for example, be indeterministic. But there is still a productive explanation of the total state in terms of the operation of the laws through time. And let's contrast this with the *Humean Package*: the idea that the total physical state of the universe (through all time) is ontologically fundamental, and that whatever laws and direction of time there might be must merely be a matter of some generic features of the Humean Mosaic. Then the first order of business has been to show that my non-Humean package really is an alternative account that runs into no obvious logical, methodological, or scientific objections.

But in the choice between the packages, my own motivation is nothing more complicated than methodological conservatism. The non-Humean package is, I think, much closer to the intuitive picture of the world that we begin our investigations with. Certainly, the fundamental asymmetry in the passage of time is inherent in our basic initial conception of the world, and the fundamental status of the laws of physics is, I think, implicit in physical practice. Both of the strands of our initial picture of the world weave together in the notion of a productive explanation, or account, of the physical universe itself. The universe, as well as all the smaller parts of it, is *made*: it is an ongoing enterprise, generated from a beginning and guided towards its future by physical law. The Humean must, at the most foundational level, reject this picture. That, of course, is no proof that the Humean is wrong: many of our most cherished initial beliefs about the world have been overturned by scientific theorizing. But I don't think that *scientific results* have, as yet, impeached the basic non-Humean picture, and no *philosophical arguments* give us reason to displace it. The metaphysics within physics is, as of now, non-Humean, and we can do no better as philosophers than embrace it.

Epilogue:
A Remark on the Method of Metaphysics

> The most exquisite folly is made of wisdom spun too fine.
> (Benjamin Franklin)

For the last quarter-century—or perhaps the last quarter-millennium—some considerable effort has been directed to the 'analysis' of a supposed metaphysical problem, which goes by the name 'The problem of homogeneous spinning disks' or 'homogeneous spinning spheres'. In its modern incarnation, the problem stems from lectures delivered by Saul Kripke and David Armstrong in 1979 and 1980 respectively. The target of the problem may be taken to be Lewis's doctrine of Humean Supervenience, or more broadly the metaphysical analysis of persistence through time. The particular form of the problem I wish to address has the following structure:

1. It is asserted that there are two metaphysically possible, distinct situations: that of a perfectly homogeneous sphere (or disk, or other rotationally symmetric object) at rest, and that of an exactly similarly shaped homogeneous object that is spinning on its axis of symmetry.
2. It is demanded that a metaphysical view—such as Lewis's Humeanism or a particular view about persistence through time—must allow for these as distinct possibilities.
3. It is argued that a particular view cannot (alternatively: can) allow for these two possibilities, and the view in question is thereby weakened (alternatively: strengthened).

It is pretty obvious how to use this case, for example, against Lewis: the two situations appear to be identical with respect to the 'Humean Mosaic' of properties that Lewis countenances, and so, by Humean Supervenience, would have to be the same in all respects.¹ So Lewis's Humean doctrine cannot account for these distinct possibilities. So Lewis must be wrong.

I have no brief to defend Lewis's Humeanism: indeed, it is wrong. Nor do I have any brief to defend any particular doctrine of 'persistence through time'. But it must be pointed out that the little argument sketched above has no value at all as a tool for doing metaphysics, and that the willingness to consider it a problem is indicative of a fundamental error of philosophical method.

The problem is simple: we have no reason whatsoever to claim to know that the two supposedly distinct, metaphysically possible situations are, either of them (much less both of them), metaphysically possible. The first premiss of the argument fails, so the whole thing can't get off the ground.

Begin with the obvious: we have every reason to believe that there is no such thing as 'perfectly homogeneous matter'. The atomic theory of matter is as well established as any scientific theory is likely to be. If by 'matter' one intends the sort of stuff that solid bodies (among other things) are composed of, then it is not homogeneous.

Second: all of our *experience* of rotating bodies is experience of actual things. Therefore, we have no experience of rotating, or non-rotating, homogeneous matter. We know, sure enough, that it is possible for solid bodies to rotate, for there are actual rotating bodies with which we are familiar. So in order to have any reason to accept that the pair of situations envisaged in the first premiss are each metaphysically possible, we would have to have reason to accept two distinct claims which our experience does not directly support: that perfectly homogeneous matter is itself metaphysically possible, and that if it were possible, it would have available to it distinct states of rotation and non-rotation. The question is: what gives us any reason to accept either of these things?

We evidently can have very good reason to believe that certain non-actual states of affairs involving material objects are metaphysically possible. The vast majority, if not the totality, of everyday instances of such a belief are instances of things taken to be *physically* possible. And states are judged to be physically possible because they are taken to be models of some (explicit or implicit)

---

[1] There is sometimes a pointless little debate over questions like: wouldn't a rotating sphere bulge around the middle and so be Humeanly distinguishable from the same sphere at rest? This is pointless because the problem does not require that the spinning object be what you would get by putting the resting object into motion. If a spinning sphere would bulge, it would still retain its symmetry about the axis of rotation. So then you compare it to a resting, similarly shaped, spheroid.

physical theory. Consider Aristotle. He surely would have accepted the possibility of rotating, or non-rotating, homogeneous spheres. For Aristotle (mistakenly) believed that actual matter was homogeneous, and (correctly) believed that actual matter can rotate. He may not have believed that any actual material object was perfectly spherical, or perfectly homogeneous, but absent some argument to the contrary he would have considered the possibility of a perfectly homogeneous sphere to be consistent with the actual nature of matter. So it is evident why Aristotle would have taken the pair of states mentioned in the main premiss to be possible: he would have taken them to be physically possible. Insofar as Aristotle was justified in his belief in his physics, he would have been justified in accepting the first premiss.

But we are evidently not justified in taking these states to be physically possible for matter. Just the opposite, we are morally certain that matter is not homogeneous. Interestingly, one could argue that contemporary physics does recognize a physical entity that can be perfectly homogeneous: a field. Perhaps the electric field around a point particle is perfectly spherically symmetric. But this does us no good, since contemporary physics does not suggest that such a spherically symmetric field could have distinct states of rotation and non-rotation. One can *mathematically* 'spin' a field by just mapping the field values at points to other points gotten by rotation, but for a field that is symmetric about the axis of rotation, the result is the very same state one started with. The field does not spin.

Nor is there any model of physics as we now have it that allows for perfectly homogeneous matter. So we have no grounds whatever to suppose that homogeneous matter is physically possible. And we therefore do not have the *usual* grounds that we have for thinking that a non-actual state is metaphysically possible.

One might object: there is a perfectly good physical theory that can handle homogeneous matter, namely continuum mechanics. So one should look to continuum mechanics to see whether the conundrum can be formulated.

But there is a confusion here about the sense in which continuum mechanics is a 'perfectly good physical theory'. Continuum mechanics can be of great use to physicists and engineers interested in particular physical problems, like turbulence and water flow. But its usefulness in these cases clearly derives from its status as a good approximation. Continuum mechanics is first a piece of mathematics, and as such has whatever mathematical structure happened to be put into it. For example, in continuum mechanics one just postulates, as a matter of specifying the mathematical structure, that the state of a continuum is to be characterized by a *velocity field*: at every point of the continuum, there is a vector called the velocity at that point. It is tempting—but from a mathematical point of view completely unnecessary—to interpret the velocity vector as indicating 'how fast, and in what direction, that bit of the continuum is moving'. But the mathematics doesn't require *any* interpretation: states of the field are given a dynamics, and evolve into later states of the field. No 'persistence conditions' for underlying 'bits' of the field are required. (The mathematics of the continuum can be used similarly to model wave phenomena in a medium, even though the 'velocity of the wave' at a point is not considered to be the velocity of any persisting bit of anything in the medium. Water waves in the ocean can be 'going in to shore' even though the water itself is basically going up and down.)

Even if one could argue that continuum mechanics presupposes some material persistence conditions for bits of homogeneous matter, still, that would play no role in explaining why continuum mechanics is a useful tool for doing physics. Again, we know that water and air are not homogeneous materials. If we can successfully use continuum mechanics to treat them, to a good approximation, it is certainly not because there must be persistence conditions for actual bits of homogeneous water. And if one *could* specify (which one can't) what would count as a 'homogeneous version of H_2O', it would be an open question whether continuum mechanics would provide a good mathematical account of its behavior. Presumably, one would have to try to work the homogeneous version of H_2O into actual physics and see what happens.

At this point, the metaphysician may try a decisive move. The metaphysician can grant everything put forward so far: there is no reason to believe, and every reason to disbelieve, that the situation described in the first premiss is physically possible. But *metaphysical* possibility extends more widely than physical possibility. It is metaphysically possible that physics might have been different from what it is, and that it might allow for completely homogeneous matter, and that the homogeneous matter it allows for might be correctly described by continuum mechanics, and that the velocity field might be correctly interpreted as giving the velocity of a persisting bit of the homogeneous matter. Then the continuum mechanics will have as models both rotating and non-rotating homogeneous spheres, and these will allow our argument to go through.

It is, I think, an important open question whether metaphysical possibility extends more widely than physical possibility. Indeed, it is an open question exactly what this first question exactly means. But let's set that aside. The obvious rejoinder to this move is that it makes metaphysics out to be nothing but *the analysis of fantastical descriptions produced by philosophers*, and, not surprisingly, these fantastical descriptions will have in them whatever features the philosophers decided to put into them.

Suppose, for example, one becomes interested in the problem of the persistence of objects. The foundation for this problem is surely the initial belief that actual material things sometimes persist. Maybe, at the end of the day, one will abandon that belief: this would constitute one sort of metaphysical resolution of the problem. No account of the 'fundamental metaphysical nature of the persistence' is required if there is no reason to think there is such a thing. (There is no metaphysical problem of the fundamental nature of Aristotle's quintessence, since there is no such thing.) Or maybe one thinks that an account can be given, and the initial belief that actual material things (at least sometimes) persist can be upheld. Well and good: that account need only be responsible for accounting for the persistence of actual material things. Probably the resources of such an account can be brought to bear to decide about how various possible situations—like actuality with respect to the features used to explicate actual persistence—would have contained persistent objects. But as soon as one wanders off to merely imaginary situations that do not share the relevant features with actuality, all bets are off. If one *stipulates* that there are certain persistence conditions in these cases, so be it: but don't pretend that such a stipulation is a metaphysical discovery. Perhaps there can be a dispute over whether the imaginary situation is *really* metaphysically possible, but if so, then one cannot appeal to the *evident* metaphysical possibility of the situation to help settle a dispute between competing metaphysical accounts, which is what the structure of the argument we began with purports to do.

So either we have some *serious* reason to think that a situation is possible (such as a demonstration that it is a model of some reasonably plausible physics, or a plausibility argument that it is a model of actual physics) or we simply think it is possible because we have produced a description and decided to nominate it a metaphysical possibility. The latter approach simply fails to make any contact with reality, and it is hard to see why discussion of such cases should be of any interest to ontology.

Let's take another case that can show some of the same features. There is a voluminous philosophical literature on the 'problem of time travel'. What exactly is this problem? There are several distinct issues in the neighborhood and they must be carefully distinguished.

One question is whether time travel can be shown to be metaphysically impossible because the very concept is somehow internally incoherent, or 'leads to paradox'. The familiar discussions of the grandfather paradox fall here. Attempts to prove the impossibility of time travel in this way have a perfectly respectable form: if a description is shown to be internally incoherent or self-contradictory, then it cannot represent any sort of a possibility. Logical impossibility is a species of metaphysical impossibility. Such attempts to prove the impossibility of time travel by internal incoherence fail, as the production of logically coherent accounts of time travel show.

But while logical impossibility is a species of metaphysical impossibility, logical possibility is not a species of metaphysical possibility. Descriptions can be logically impeccable, such as 'Cicero was not Tully', yet describe metaphysically impossible situations. Or at least: any attempt to argue that logical and metaphysical possibility coincide runs into a host of apparent counter-examples, produced both by the familiar Kripke/Putnam examples and by the impossibility of reducing mathematics to logic. Most parties to metaphysical debates would not identify logical and metaphysical possibility.

A second line of discussion about time travel attempts not to prove it to be impossible but rather to prove it to be metaphysically possible. The most direct way to do this would evidently be to prove it to be actual. Unfortunately, philosophers have not made much progress in this direction. The next best thing would be to prove it to be physically possible, but here a thicket of problems arises. The notion of physical possibility—which has quite uncontroversial application in other cases—has no uncontroversial application here.

To explicate the notion of physical possibility, we earlier spoke of the models of a physical theory. And if one specifies exactly what constitutes a particular physical theory, this is unproblematic. So, for example, if one specifies that all one means is a model of Einstein's field equations, then it is unproblematic to show that among these models are models with closed timelike curves: time travel in the usual sense. If these field equations had no such models, and if we thought that these equations were the correct equations for the world, then the physical impossibility of time travel would be secured. This would still leave open the question of the metaphysical possibility of time travel, if metaphysical possibility is wider than physical, but let's leave that aside.

As it turns out, the Einstein field equations do, by themselves, allow for time travel. This is sometimes because the global topology of these models is put in by hand: the field equations are local and do not specify the global topology of the manifold.[2] But even so, physicists who work in the field were not automatically convinced that this shows time travel to be physically possible. Indeed, the attitude of the main developers of the field in the 1970s was just the opposite: that the existence of closed timelike curves in a solution indicated that it was not a *physical* solution. It is notable in this case that the equations do not force the existence of closed timelike curves in this sense: for any initial condition one can specify, there is a global solution for that initial condition that does not have closed timelike curves.

So suppose we agree that time travel is non-actual. There is no straightforward sense in which physics—or more precisely, the sort of physics encapsulated in the Einstein field equations—could force one to accept that time travel is physically possible. Indeed, it is not clear whether in any decent sense the field equations, by themselves, suggest that it is possible. For one is free to maintain that it is in the fundamental metaphysical nature of time itself that closed timelike curves are not possible, that moving forward in time from any given event cannot bring one back to that event.

But surely this is a bold supposition: on what possible grounds could one maintain such insight into the very nature of time itself, to see that time will not abide this sort of recurrence? *How could one know* that the basic nature of time itself is incompatible with time travel?

Quite so: it would be bold to claim such insight into the metaphysical nature of time, especially given the existence of mathematical models of some plausible laws of physics that permit time travel. But is it not equally bold to claim insight into the nature of time that shows time travel to be possible *if we grant that it is not actual and also that the laws of physics, operating from conditions we take to be possible, do not require it*. If we grant that it does not happen, and that relative to conditions we agree to be possible it need not happen, by what right do we accept that it could happen?

The case of time travel, then, ought to be treated as a case of spoils to the victor. We have, as yet, no evidence for the actuality of closed timelike curves. We have neither direct evidence (such as a time machine) nor indirect evidence (such as a well-supported fundamental physical theory that implies the existence of such curves in the actual world). Philosophers, for many centuries, have had the strong intuition that time travel is not possible: one cannot, by going forward in time, get back to the past. More recently,

---

[2] Not all examples of closed timelike curves result from a non-simple global topology. Gödel's famous solution, for example, uses a space-time that is topologically simple.

some philosophers have developed the opposite intuition: barring any direct argument to the contrary, one ought to suppose that time travel *is* possible. But any such 'intuitions' about the fundamental nature of time itself appear to be completely ungrounded. Our experience of time is experience of actual time. If we agree that actual time contains no such time travel, why should we repose any confidence in our 'intuitions' about what could or could not happen? Maybe one's *concept* of time clearly allows, or disallows, time travel, but unfortunately time itself need not be accommodating to our concepts. Aristotle's *concept* of matter allowed for homogeneous rotating disks, but actual disks have not been kind to him.

Just as a supposition that the rotating and non-rotating homogeneous disks are metaphysically possible should not be available as a *premiss* from which one can draw metaphysical *conclusions*, so the supposition of the possibility, or of the impossibility, of time travel ought not to be treated as a datum to which a theory of time must be held responsible. We might be very lucky: time travel could be shown to be possible by showing it to be actual. We are not likely to be lucky in this way about the disks. Or again, maybe we will come to accept a physics that either rules out, or obviously permits, time travel or the disks. If so, this will be a fortuitous by-product: our grounds for accepting the physics will not be either that it does, or does not, allow for these possibilities. Or the physics we end up accepting may not, in any obvious way, settle the issue. We would then have to engage in a serious consideration of whether there is any fact—physical or metaphysical—of which we are left ignorant.

References

Albert, D. Z (1992), *Quantum Mechanics and Experience* (Cambridge, Mass.: Harvard University Press).
____ (2000), *Time and Chance* (Cambridge, Mass.: Harvard University Press).
Armstrong, D. M. (1978), *Universals and Scientific Realism* (Cambridge: Cambridge University Press).
____ (1983), *What is a Law of Nature?* (Cambridge: Cambridge University Press).
Barbour, J. (2000), *The End of Time* (Oxford: Oxford University Press).
Bell, J. S. (1987), *Speakable and Unspeakable in Quantum Mechanics* (Cambridge: Cambridge University Press).
Bohm, D., and Hiley, B. J. (1993), *The Undivided Universe* (London: Routledge).
Born, M. (1971), *The Born–Einstein Letters*, trans. I. Born (New York: Walker).
Carnap, R. (1959), 'The Elimination of Metaphysics through Logical Analysis of Language', in A. J. Ayer (ed.), *Logical Positivism* (Glencoe, Ill.: Free Press).
Carroll, J. (1994), *Laws of Nature* (Cambridge: Cambridge University Press).
Collins, J. (2000), 'Pre-emptive Prevention', *Journal of Philosophy*, 97: 223–34.
____ Hall, N., and Paul, L. A. (2004), *Causation and Counterfactuals* (Cambridge, Mass.: MIT Press).
Comins, N. (1995), *What If the Moon Didn't Exist?* (New York: HarperPerennial).
Dretske, F. (1977), 'Laws of Nature', *Philosophy of Science*, 44: 248–68.
Earman, J. (1974), 'An Attempt to Add a Little Direction to "The Problem of the Direction of Time"', *Philosophy of Science*, 41: 15–47.
____ (1984), 'Laws of Nature: The Empiricist Challenge', in R. J. Bogdan (ed.), *D. M. Armstrong* (Dordrecht: Reidel).
____ (1986), *A Primer of Determinism* (Dordrecht: Reidel).
Galilei, Galileo (1974), *Two New Sciences* (Madison: University of Wisconsin Press).
Ghirardi, G. C., Rimini, A., and Weber, T. (1986), 'Unified Dynamics for Microscopic and Macroscopic Physics', *Physical Review*, D 34: 470–91.
Glymour, C. (1980), *Theory and Evidence* (Princeton: Princeton University Press).
Gödel, K. (1949), 'A Remark about the Relationship between Relativity Theory and the Idealistic Philosophy', in P. Schilpp (ed.), *Albert Einstein: Philosopher-Scientist* (La Salle, Ill.: Open Court).
Goodman, N. (1983), *Fact, Fiction and Forecast* (Cambridge, Mass.: Harvard University Press).
Guth, A., and Steinhardt, P. (1989), 'The Inflationary Universe', in P. Davies (ed.), *The New Physics* (Cambridge: Cambridge University Press).

Hall, N. (2004), 'Two Concepts of Causation', in Collins, Hall, and Paul 2004.
Hawking, S., and Ellis, G. (1973), *The Large Scale Structure of Space-Time* (Cambridge: Cambridge University Press).
Horwich, P. (1987), *Asymmetries in Time* (Cambridge, Mass.: MIT Press).
Lewis, D. (1973a), *Counterfactuals* (Cambridge, Mass.: Harvard University Press).
―― (1973b), 'Causation', *Journal of Philosophy*, 70: 556–67. Reprinted in Lewis 1986a, pp. 159–213.
―― (1986a), *Philosophical Papers*, Vol. ii (Oxford: Oxford University Press).
―― (1986b), *On the Plurality of Worlds* (Oxford: Basil Blackwell).
Loewer, B. (1996), 'Humean Supervenience', *Philosophical Topics*, 24: 101–27.
McDermott, M. (1995), 'Redundant Causation', *British Journal for the Philosophy of Science*, 46: 523–44.
McTaggart, J. M. E. (1927), *The Nature of Existence* (Cambridge: Cambridge University Press).
Maudlin, T. (1993), 'Buckets of Water and Waves of Space: Why Space-Time is Probably a Substance', *Philosophy of Science*, 60: 183–203.
―― (1994), *Quantum Non-Locality and Relativity* (Oxford: Basil Blackwell).
―― (1996), 'Space-Time in the Quantum World', in J. Cushing, A. Fine, and S. Goldstein (eds.) *Bohmian Mechanics and Quantum Theory: An Appraisal* (Dordrecht: Kluwer).
―― (2001), 'Remarks on the Passing of Time', *Proceedings of the Aristotelian Society*, 102 (part 3): 237–52.
―― (2007), 'What Could Be Objective about Probability?', forthcoming in *Studies in the History and Philosophy of Science*.
Mermin, N. D. (1990), *Boojums All the Way Through* (Cambridge: Cambridge University Press).
Misner, C., Thorne, K., and Wheeler, J. A. (1973), *Gravitation* (New York: W. H. Freeman).
Newton, I. (1966), *Principia* (Berkeley and Los Angeles: University of California Press).
Penrose, R. (2005), *The Road to Reality* (New York: Knopf).
Perle, P. (1990), 'Toward a Relativistic Theory of Statevector Reduction', in A. I. Miller (ed.), *Sixty-Two Years of Uncertainty* (New York: Plenum Publishing).
Price, H. (1996), *Time's Arrow and Archimedes'' Point* (Oxford: Oxford University Press).
Quine, W. V. O. (1951), 'Two Dogmas of Empiricism', *Philosophical Review*, 60: 20–43.
―― (1953), 'On What There Is', in *From a Logical Point of View* (New York: Harper).
Redhead, M. (1987), *Incompleteness, Nonlocality and Realism: A Prolegomenon to the Philosophy of Quantum Mechanics* (Oxford: Clarendon Press).

Reichenbach, H. (1958), *The Philosophy of Space and Time* (New York: Dover).
Russell, B. (1923), 'Vagueness', *Australasian Journal of Philosophy and Psychology*, 1: 84–92.
Schaffer, Jonathan (2000), 'Causation by Disconnection', *Philosophy of Science*, 67: 285–300.
Skow, B. (forthcoming), 'What Makes Time Different from Space?', *Nous*.
Tooley, M. (1977), 'The Nature of Law', *Canadian Journal of Philosophy*, 7: 667–98.
Vallentyne, P. (1988), 'Explicating Lawhood', *Philosophy of Science*, 55: 598–613.
van Fraassen, B. (1980), *The Scientific Image* (Oxford: Clarendon).
—— (1989), *Laws and Symmetry* (Oxford: Clarendon).
Williams, D. C. (1951), 'The Myth of Passage', *Journal of Philosophy*, 48. Page numbers given are for the reprinted version in R. Gale (ed.), *The Philosophy of Time* (Garden City, NY: Anchor).

Index

affine connection, 90–2
Albert, David, vii, 64, 118–19, 127, 142, 176–7
Aristotle, 79–80, 101, 186, 188, 191
Armstrong, David, 2, 10–11, 14, 17, 34, 87, 102, 186
Arntzenius, Frank, vii, 142

Barbour, Julian, 88
base space, 94–102
Bell, John Stewart, 49, 62–4, 104, 117, 192
Bell's inequality, 126
Bell's theorem, 49, 62
Best Systems account of law. *See* Mill–Ramsey–Lewis account of law
Bishop, Elizabeth, 50
block universe, 109–11, 115, 121
Bohm's theory, 57
Born, Max, 53
Buffon's needle problem, 68

Carnap, Rudoph, 21, 69–71, 81
Carroll, John, 5, 52
Cauchy surface, 18–23, 28–33, 48
causation
    metaphysical structure of, 167–8
    proximate, 164
    remote, 164–7
ceteris paribus clause, 24, 29–33, 45, 180
chance, 68
Collins, John, 146, 161
Comins, Neil, 67
continuum mechanics, 186
Conway, John, 149
counterfactuals, 21–34, 65
    and causation, 144–9
    and stochastic laws, 28, 49
    support of by laws, 7–8
    three-step process for evaluating, 23
covering law model of explanation, 8, 69, 172–3
CPT theorem, 117–18

dispositional fact, 72
distance relation, not metaphysically pure, 87
Doppelgänger, time-reversed, 121–5
Dretske, Fred, 10, 34
duplication, Lewis's notion of, 103

Earman, John, 18, 52–3, 71–5, 108–9, 116, 127, 142
Eberhard, Phillippe, 49
Einstein, Albert, 7, 53–4, 63–4, 115
Elga, Adam, 142
entangled states, quantum mechanical, 56
entropy, 3, 117, 122, 124, 128–38, 142, 176
    and direction of time, 128
explanation,
    and direction of time, 130–4
    and laws, 8, 34–40

fiber bundle, 94–106
Fitzgerald, F. Scott, 122
FLOTE. *See* Fundamental Law of Temporal Evolution
Franklin, Benjamin, 184
Fundamental Law of Temporal Evolution (FLOTE), 12–22, 34–5, 47–8, 172–4, 178

gauge theory, 2, 80, 89, 94–6
General Relativity, 64, 67, 117, 189–90
Ghirardi, GianCarlo, 62
Ghirardi, Rimini and Weber, spontanious collapse theory of, 62
Glymour, Clark, 4
Gödel, Kurt, 115–16, 126, 190
Gold universe, 131
Goldstein, Sheldon, vii, 142
Goodman, Nelson, 26, 106
Goosens, William, 146
Guth, Alan, 40–4

Hall, Ned, vii, 142, 144, 164, 174,
Heisenberg Uncertainty Principle, 55
Hempel, Carl, 105
Higgs field, 42–3, 138
Horwich, Paul, 117–18, 127
Hume counterfactual, 143–4, 161, 164–6
Hume, David, 4, 20, 50, 68–71, 79, 103–4, 143–154, 167, 171, 181
Humean Supervenience, 1–4, 50, 53, 61–3, 71, 75–6, 105, 170, 184

ideal gas law, 13
inflationary cosmology, 40–4
initial conditions, 9, 16–17, 36, 39–44, 66, 130
in-your-face opponent, 126–7

Kant, Immanuel, 78, 115, 126–7
Kaon Orientation, 120, 135, 137
Kripke, Saul, 184, 189

Law of Temporal Evolution (LOTE), 13–14, 31
laws
  and causation, 151
  and explanation, 34–40
  and physical possibility, 18
  as relations between universals, 10–11
  epistemological access to, 17
  logical form of, 10–12
  of coexistence, 13
  of nature, characterization of, 36
  of special sciences, 158–61
  quasi-Newtonian, 155–6
  stochastic, 16, 28, 49, 68
  time reversal invariance of, 118–19
Lewis, David, 1–4, 6, 9, 15–16, 19–23, 32–4, 50–2, 61–8, 71, 75–7, 81, 84–5, 93, 103, 105, 143, 146, 150, 154, 170–4, 180, 184–5
local stochastic process, 20
Loewer, Barry, v, 61–2, 170–3
LOTE. *See* Law of Temporal Evolution

Many Minds theory, 62
Maudlin, Clio, vii
Maudlin, Maxwell, vii

Maudlin, Vishnya, vii
Maudlin's Non-Humean Package, 182
McCall, Storrs, 34
McDermott, Michael, 161–4
McLaughlin, Brian, 45
McTaggart, J. M. E., 126
Mermin, David, 29–31
metaphysical purity, definition of, 86
metaphysics, nature of, 78, 104
Mill, John Stewart, 9, 15, 34, 105, 150, 172
Mill–Ramsey–Lewis account of law, 9, 15, 150, 172
mind-body problem, 107
Minkowski space-time, 67, 138
moving now, 110

necessity, nomic, 21
neuron diagrams, 160
Newton, Isaac, 8–9, 48, 67, 110, 129, 148, 155–6
Newton's equations of motion, 11, 119, 155
Noether's theorem, 37
nomic structure, 34
nominalism, Quinean, 83
non-separability, quantum, 62

occurrent fact, 72–5
Ockham's Razor, 3–4, 76, 105, 120, 129–30, 179–1
orientation, on a Relativistic space-time, 116, 138

parallel transport, 91–3, 96, 100
parallelism, not metaphysically pure, 90
Parmenides, 115, 126–7
passage of time
  elucidation of concept, 108
  rate of, 112
Past Hypothesis, 176–7
Penrose, Roger, 97, 100
Perle, Philip, 62
philosophical method, bad, 2, 146–7, 184–91
physical statism, 51–3, 63–4, 67
physicalism, 51–2

Index

possibility
  logical, 189
  metaphysical, 185, 188
  physical, 18
Price, Huw, 109–15, 121, 124–7, 131, 142
principle of superposition, 54–6
probabilistic laws. *See* stochastic laws
product states, 56–7
property
  abundant, 84
  dipositional, 72
  occurent, 72
  sparse, 84
protected window, 162–4
Putnam, Hilary, 189

quantum non-locality, 19–20
quark color, 53, 75–6, 94–96
Quine, W. V. O., 80–5, 104
Quine's recipe, 81

Ramsey, Frank, 9, 15, 34, 105, 150, 172
Redhead. Michael, 28–31, 49
regularity account of law, 9
Reichenbach, Hans, 6
Relativity, theory of, 7, 115
Rimini, Alberto, 62
rotating spheres, homogeneous, 184–8
Russell, Bertrand, 79, 126

Sayre-McCord, Geoff, 103
Schaffer, Jonathan, 164
schlag, good for ontology, 4
Schrödinger's Equation, 11
second law of thermodynamics, 129
section of a fiber bundle, definition of, 98
separability, 51–8, 61–4, 79

singlet state, 56–61
Sklar, Lawrence, 127
Skow, Brad, 139
SLOTE. *See* Special Law of Temporal Evolution
Socrates, blood type of, 73
space-time diagram, 139
spatializing time, 140
Special Law of Temporal Evolution (SLOTE), 45–7
Special Relativity, 67–8
special sciences, 158–61
Stapp, Henry, 49
Steinhardt, Paul, 40–4
stochastic laws, 16, 28, 49, 68
superposition, 55, 59

tangent bundle, 94, 100
temporalizing space, 140
time reversal invariance, 118–19
time reversal operation, 119
time travel, 175, 188–91
time-reversed Doppelgänger, 121–4
Tooley, Michael, 10, 34
triangle Inequality, 88–9, 99–101
triplet state, 56–61
tropes, 80, 85–6, 93–7, 101–3

universals, 1–2, 11, 14–17, 80, 83–6, 93–7, 101–3

Vallentyne, Peter, 34–9
van Fraassen, Bas, 8–11, 15, 34, 37–9, 44, 173

Weber, Tulio, 62
Weslake, Brad, 142
Wheeler, John, 12
Williams, D. C., 121–8

Made in the USA
Lexington, KY
12 December 2011